# PhysioEx version 2.0 Laboratory Simulations in Physiology

Peter Z. Zao, North Idaho College

An Imprint of Addison Wesley Longman, Inc.

San Francisco • Reading, Massachusetts • New York • Harlow, England
Don Mills, Ontario • Sydney • Mexico City • Madrid • Amsterdam

Publisher: Daryl Fox
Senior Project Editor: Lauren Fogel
Managing Editor: Wendy Earl
Production Supervisors: Janet Vail and Vivian McDougal
Art and Design Supervisor: Bradley Burch
Text Designer: Juan Vargas
Cover Designer: Lillian Carr

Copyright © 2000 by Addison Wesley Longman, Inc. This title published under the Benjamin/Cummings imprint.

All rights reserved. No part of this publication may be reproduced, stored in a retrieval system, or transmitted, in any form or by any means, electronic, mechanical, photocopying, recording, or any other media or embodiments now known or hereafter to become known, without the prior written permission of the publisher. Manufactured in the United States of America. Published simultaneously in Canada.

Many of the designations used by manufacturers and sellers to distinguish their products are claimed as trademarks. Where those designations appear in this book, and the publisher was aware of a trademark claim, the designations have been printed in initial caps or all caps.

The Author and Publisher believe that the lab experiments described in this publication, when conducted in conformity with the safety precautions described herein and according to the school's laboratory safety procedures, are reasonably safe for the students to whom this manual is directed. Nonetheless, many of the described experiments are accompanied by some degree of risk, including human error, the failure or misuse of laboratory or electrical equipment, mismeasurement, spills of chemicals, and exposure to sharp objects, heat, bodily fluids, blood, or other biologics. The Author and Publisher disclaim any liability arising from such risks in connection with any of the experiments contained in this manual. If students have any questions or problems with materials, procedures, or instructions on any experiment, they should *always* ask their instructor for help before proceeding.

ISBN 0-8053-6168-5

4 5 6 7 8 9 10—CRS—03 02 01 00

Credits appear on page iv.

Addison Wesley Longman, Inc.
1301 Sansome Street
San Francisco, CA 94111

## The Benjamin/Cummings Series in Human Anatomy and Physiology

**Textbooks**

By R.A. Chase
*The Bassett Atlas of Human Anatomy* (1989)

By Kapit/Elson
*The Anatomy Coloring Book*, second edition (1993)

By Kapit/Macey/Meisami
*The Physiology Coloring Book,* second edition (1999)

By E.N. Marieb
*Human Anatomy and Physiology*, fourth edition (1998)

*Human Anatomy and Physiology, Study Guide*, fourth edition (1998)

*Human Anatomy and Physiology Laboratory Manual, Cat Version*, sixth edition (2000)

*Human Anatomy and Physiology Laboratory Manual, Fetal Pig Version*, sixth edition (2000)

*Human Anatomy and Physiology Laboratory Manual, Main Version*, fifth edition (2000)

*Essentials of Human Anatomy and Physiology*, fifth edition (1997)

*The A&P Coloring Workbook: A Complete Study Guide*, fifth edition (1997)

By E.N. Marieb and J. Mallatt
*Human Anatomy*, second edition (1997)

# Contents

**Preface**   v

Exercise 5B    The Cell—Transport Mechanisms
and Cell Permeability: Computer
Simulation   1

Exercise 16B   Skeletal Muscle Physiology:
Computer Simulation   13

Exercise 33    Cardiovascular Dynamics:
Computer Simulation   25

Exercise 35B   Frog Cardiovascular Physiology:
Computer Simulation   37

Exercise 40B   Chemical and Physical Processes
of Digestion: Computer
Simulation   47

Exercise 48B   Respiratory System Mechanics:
Computer Simulation   59

Exercise 49B   Renal Physiology—The Function
of the Nephron: Computer
Simulation   67

Using the Histology Instructions       75

Review Sheet for 5B         77

Review Sheet for 16B        81

Review Sheet for 33         85

Review Sheet for 35B        89

Review Sheet for 40B        91

Review Sheet for 48B        95

Review Sheet for 49B        99

# Credits

## ILLUSTRATIONS

**Exercise 5**
5B.1–5B.5: Steve McEntee.

**Exercise 16**
16B.1–16B.4: Steve McEntee.

**Exercise 33**
33.1 and 33.2: Steve McEntee.

**Exercise 35**
35B.1–35B.3: Steve McEntee.

**Exercise 40**
40B.1–40B.3: Steve McEntee.

**Exercise 48**
48B.1–48B.2: Steve McEntee.

**Exercise 49**
49B.1–49B.2: Steve McEntee.

# Preface

PhysioEx version 2.0 CD-ROM is an exciting multimedia product created for the human physiology laboratory. Unlike the typical tutorial-based computer supplements, the seven easy-to-use physiology experiments and the histology tutorial on PhysioEx version 2.0 allow students to freely investigate various topics in physiology electronically, and yet still have the structure and security of a written lab exercise guiding them through their process of discovery. Students can conduct or review the experiments at home on their personal computer or in any campus computer center. PhysioEx version 2.0 provides convenient "laboratory access" to students enrolled in Internet-based distance education courses.

## PhysioEx version 2.0 topics include

Exercise 5B, Cell Transport Mechanisms and Permeability. Explores how substances cross the cell's membrane. Simple and facilitated diffusion, osmosis, filtration, and active transport are covered.

Exercise 16B, Skeletal Muscle Physiology. Provides insights into the complex physiology of skeletal muscle. Electrical stimulation, isometric contractions, and isotonic contractions are investigated.

Exercise 33, Cardiovascular Dynamics. Allows students to perform experiments that would be difficult if not impossible to do in a traditional laboratory. Topics of inquiry include vessel resistance and pump (heart) mechanics.

Exercise 35B, Frog Cardiovascular Physiology. Variables influencing heart activity are examined. Topics include setting up and recording baseline heart activity, the refractory period of cardiac muscle, and an investigation of physical and chemical factors modifying heart rate.

Exercise 40B, Chemical and Physical Processes of Digestion. Turns the student's computer into a virtual chemistry lab where enzymes, reagents, and incubation conditions can be manipulated (in compressed time) to examine factors that affect enzyme activity.

Exercise 48B, Respiratory System Mechanics. Investigates physical and chemical aspects of pulmonary function. Students will collect data simulating normal lung volumes. Other activities examine factors such as airway resistance and the effect of surfactant on lung function.

Exercise 49B, Renal System Physiology. Simulates the function of a single nephron. Topics include factors influencing glomerular filtration, the effect of hormones on urine formation, and glucose transport maximum.

# The Cell—Transport Mechanisms and Cell Permeability: Computer Simulation

## Objectives

1. To define *differential permeability; diffusion (simple diffusion, facilitated diffusion,* and *osmosis); isotonic, hypotonic,* and *hypertonic solutions; passive transport; active transport; pinocytosis; phagocytosis;* and *solute pump.*

2. To describe the processes that account for the movement of substances across the plasma membrane and to indicate the driving force for each.

3. To determine which way substances will move passively through a differentially permeable membrane (given the appropriate information on concentration differences).

## Materials

Part I—Demonstrations

❏ Millimeter rulers

*1. Diffusion of a dye through agar*

Prepared the morning of the laboratory session with setup time noted. A crystal of potassium permanganate dye is placed gently in a petri plate containing 12 ml of 1.5% agar gel. An equal amount of methylene blue crystals is placed on the agar surface approximately 10 centimeters away from the potassium permanganate crystal.

*2. Osmometer*

Just before the laboratory begins, the broad end of a thistle tube is closed with a differentially permeable dialysis membrane, and the tube is secured to a ring stand. Molasses is added to approximately 5 cm above the thistle tube bulb and the bulb is immersed in a beaker of distilled water. At the beginning of the lab session, the level of the molasses in the tube is marked with a wax pencil.

Part II—Cell transport simulation

Minimum computer equipment required:

**Windows:**

❏ 486/66 MHz or better recommended

❏ Windows 95/98 recommended

❏ SVGA display (256 colors at 640 × 480)

❏ Minimum 16 MB available RAM

❏ Double speed CD-ROM drive (quad-speed or higher recommended)

❏ Sound card

❏ Speakers or headphones

❏ Printer

**Macintosh:**

❏ 68040 processor or Power Macintosh recommended

❏ System 7.1 or higher

❏ Minimum 16 MB available RAM

❏ 13-inch or larger color monitor (640 × 480 resolution)

❏ Double speed CD-ROM drive (quad-speed or higher recommended)

❏ Printer

**Software:**

❏ Benjamin/Cummings PhysioEx CD-ROM (Cell Transport Mechanisms module)

The molecular composition of the plasma membrane allows it to be selective about what passes through it. It allows nutrients to enter the cell but keeps out undesirable substances. By the same token, valuable cell proteins and other substances are kept within the cell, and ex-creta or wastes pass to the exterior. This property is known as **differential,** or **selective, permeability.** Transport through the plasma membrane occurs in two basic ways. In **active transport,** the cell provides energy (ATP) to power the transport process. In the other, **passive transport,** the transport process is driven by concentration or pressure differences between the interior and exterior of the cell.

# Passive Transport

The two key passive processes of membrane transport are diffusion and filtration. Diffusion is an important transport process for every cell in the body. By contrast, filtration usually occurs only across capillary walls. Each of these will be considered in turn.

## Diffusion

Recall that all molecules possess *kinetic energy* and are in constant motion. As molecules move about randomly at high speeds, they collide and ricochet off one another, changing direction with each collision. In general, the smaller the particle, the more kinetic energy it has and the faster it moves.

When a **concentration gradient** (difference in concentration) exists, the net effect of this random molecular movement is that the molecules eventually become evenly distributed throughout the environment, i.e., the process called diffusion occurs. Hence, **diffusion** is the movement of molecules from a region of their higher concentration to a region of their lower concentration. Diffusion's driving force is the kinetic energy of the molecules themselves.

There are many examples of diffusion in nonliving systems. For example, if you uncorked a bottle of ether at the front of the laboratory, very shortly thereafter you would be sleepily nodding off as the ether molecules distributed throughout the room. Similarly, the ability to smell a friend's cologne shortly after he or she has entered the room is an example of diffusion.

The diffusion of particles into and out of cells is modified by the plasma membrane, which constitutes a physical barrier. In general, molecules diffuse passively through the plasma membrane if they are small enough to pass through its pores (and are aided by an electrical gradient), or if they can dissolve in the lipid portion of the membrane as in the case of $CO_2$ and $O_2$. The diffusion of solute particles dissolved in water through a semipermeable membrane is called **simple diffusion.** The diffusion of water through a semipermeable membrane is called **osmosis.** Both simple diffusion and osmosis involve movement of a substance from an area of its higher concentration to one of its lower concentration, i.e., down its concentration gradient.

**Diffusion of a Dye Through Agar**  Make a mental note to yourself to go to the demonstration areas near the end of the laboratory session to observe the extent of diffusion of different dyes. At that time, follow the directions for the two activities provided at the end of the diffusion simulations.

# Solute Transport Through Nonliving Membranes

This computerized simulation provides information on the passage of water and solutes through semipermeable membranes, which may be applied to the study of transport mechanisms in living membrane-bound cells.

## Getting Started

Begin by making sure you have the computer equipment and software listed in the Materials section on p. 1.

1.   Quit all applications currently running on your computer.

2.   Insert the PhysioEx CD-ROM into the CD-ROM drive and keep PhysioEx in the drive during the entire time you use the program.

3.   Follow the instructions below for your computer type.

**Windows 95/Windows 98**
The program will launch automatically after you load the CD-ROM into the drive.
If autorun is disabled on your machine follow these instructions:
1.   Double-click the My Computer icon on the Windows desktop.
2.   Double-click the PhysioEx CD icon.
3.   Look for the icon that looks like a tiny movie projector and double-click it.

**Macintosh**
Double-click the PhysioEx CD icon that appears in the window on your desktop to launch the program.

### A c t i v i t y :
## Simulating Dialysis

Choose Cell Transport Mechanisms and Permeability from the main menu. The opening screen will appear in a few seconds (Figure 5B.1). The primary features on the screen when the program starts are a pair of glass beakers perched atop a solutions dispenser, a dialysis membranes cabinet at the right side of the screen, and a data collection unit at the bottom of the display.

The beakers are joined by a membrane holder, which can be equipped with any of the dialysis membranes from the cabinet. Each membrane is represented by a thin colored line suspended in a gray supporting frame. The solute concentration of dispensed solutions is displayed at the side of each beaker. As you work through the experiments, keep in mind that membranes are three-dimensional; thus what appears as a slender line is actually the edge of a membrane sheet.

The solutions you can dispense are listed beneath each beaker. You can choose more than one solution, and the amount to be dispensed is controlled by clicking (+) to increase concentration, or (−) to decrease concentration. The chosen solutions are then delivered to their beaker by clicking the **Dispense** button on the same side. Clicking the **Start** button opens the membrane holder and begins the experiment. To clean the beakers and prepare them for the next run, click **Flush.** Clicking **Pause** and then **Flush** during a run stops the experiment and prepares the beakers for another run. You can adjust the timer for any interval between 5 and 300; the elapsed time is shown in the small window to the right of the timer.

To move dialysis membranes from the cabinet to the membrane holder, click and hold the mouse on the selected membrane, drag it into position between the beakers, and then release the mouse button to drop it into place. Each membrane possesses a different molecular weight cutoff (MWCO), indicated by the number below it. You can think of

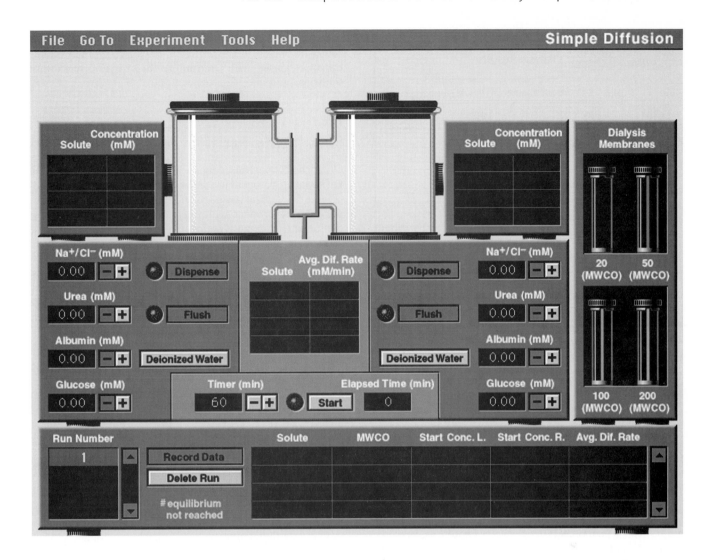

**Figure 5B.1  Opening screen of the Simple Diffusion experiment.**

MWCO in terms of pore size; the larger the MWCO number, the larger the pores in the membrane.

The Run Number window in the data collection unit at the bottom of the screen displays each experimental trial (run). When you click the **Record Data** button, your data is recorded in the computer's memory and is displayed in the data grid at the bottom of the screen. Data displayed in the data grid include the solute (Solute) and membrane (MWCO) used in a run, the starting concentrations in the left and right beakers (Start Conc. L and Start Conc. R), and the average diffusion rate (Avg. Dif. Rate). If you are not satisfied with a run, you can click **Delete Run.**

1.  Click and hold the mouse on the 20 MWCO membrane and drag it to the membrane holder between the beakers. Release the mouse button to lock the membrane into place.

2.  Now increase the NaCl concentration to be dispensed by clicking the (+) button under the left beaker until the display window reads 9.00 m*M*. Click **Dispense** to fill the left beaker with 9.00 m*M* NaCl solution.

3.  Click the **Deionized Water** button under the right beaker and then click **Dispense** to fill the right beaker with deionized water.

4.  Adjust the timer to 60 min (compressed time), then click the **Start** button. When Start is clicked, the barrier between the beakers descends, allowing the solutions in each beaker to have access to the dialysis membrane separating them. Notice that the Start button becomes a Pause button that allows you to momentarily halt the progress of the experiment so you can see instantaneous diffusion or transport rates.

5.  Watch the concentration windows at the side of each beaker for any activity. A level above zero in NaCl concentration in the right beaker indicates that Na$^+$ and Cl$^-$ ions are diffusing from the left into the right beaker through the semipermeable dialysis membrane. Record your results (+ for diffusion, − for no diffusion) in Chart 1 on p. 42. Click the **Record Data** button to keep your data in the computer's memory.

6.  Click the 20 MWCO membrane (in the membrane holder) again to automatically return it to the membranes cabinet and then click **Flush** beneath each beaker to prepare for the next run.

7.  Drag the next membrane (50 MWCO) to the holder and repeat steps 2 through 6. Continue the runs until you have tested all four membranes. (Remember: click **Flush** beneath each beaker between runs.)

## Chart 1    Dialysis Results

| Solute | Membrane (MWCO) | | | |
|---|---|---|---|---|
| | 30 | 50 | 100 | 200 |
| NaCl | | | | |
| Urea | | | | |
| Albumin | | | | |
| Glucose | | | | |

8. Now perform the same experiment for urea, albumin, and glucose by repeating steps 1 through 7 three times. In step 2 you will be dispensing first urea, then albumin, and finally glucose, instead of NaCl.

Which solute(s) were able to diffuse into the right beaker from the left?

_____

_____

Which solute(s) did not diffuse?

_____

_____

If the solution in the left beaker contained both urea and albumin, which membrane(s) could you choose to selectively remove the urea from the solution in the left beaker? How would you carry out this experiment?

_____

_____

_____

Assume that the solution in the left beaker contained NaCl in addition to the urea and albumin. How could you set up an experiment so that you removed the urea, but left the NaCl concentration unchanged?

_____

_____

_____ ■

## Facilitated Diffusion

Some molecules are lipid insoluble or too large to pass through plasma membrane pores; instead, they pass through the membrane by a passive transport process called **facilitated diffusion.** In this form of transport, solutes combine with protein carrier molecules in the membrane and are then transported *along* or *down* their concentration gradient. Because facilitated diffusion relies on carrier proteins, solute transport varies with the number of available membrane transport proteins.

### A c t i v i t y :
## Simulating Facilitated Diffusion

Click the **Experiment** menu and then choose **Facilitated Diffusion.** The opening screen will appear in a few seconds (Figure 5B.2). The basic screen layout is similar to that of the previous experiment with only a few modifications to the equipment. You will notice that only NaCl and glucose solutes are available in this experiment, and you will see a Membrane Builder on the right side of the screen.

The (+) and (−) buttons underneath each beaker adjust solute concentration in the solutions to be delivered into each beaker. Similarly, the buttons in the Membrane Builder allow you to control the number of carrier proteins implanted in the membrane when you click the **Build Membrane** button.

In this experiment, you will investigate how glucose transport is affected by the number of available carrier molecules.

1. The Glucose Carriers window in the Membrane Builder should read 500. If not, adjust to 500 by using the (+) or (−) button.

2. Now click **Build Membrane** to insert 500 glucose carrier proteins into the membrane. You should see the membrane appear as a slender line encased in a support structure within the Membrane Builder. Remember that we are looking at the edge of a three-dimensional membrane.

3. Click on the membrane and hold the mouse button down as you drag the membrane to the membrane holder between the beakers. Release the mouse to lock the membrane into place.

4. Adjust the glucose concentration to be delivered to the left beaker to 2.00 m$M$ by clicking the (+) button next to the glucose window until it reads 2.00.

5. To fill the left beaker with the glucose solution, click the **Dispense** button just below the left beaker.

6. Click the **Deionized Water** button below the right beaker, and then click the **Dispense** button. The right beaker will fill with deionized water.

7. Set the timer to 60 min and click **Start.** Watch the concentration windows next to the beakers. When the 60 minutes have elapsed, click the **Record Data** button to display glucose transport rate information in the grid at the lower edge of the screen. Record the glucose transport rate in Chart 2 on p. 43.

8. Click the **Flush** button beneath each beaker to remove any residual solution.

9. Click the membrane support to return it to the Membrane Builder. Increase the glucose carriers and repeat steps 2 through 8 using membranes with 700 and then 900 glucose carrier proteins. Record your results in Chart 2 each time.

10. Repeat steps 1 through 9 at 8.00 m$M$ glucose concentration. Record your results in Chart 2.

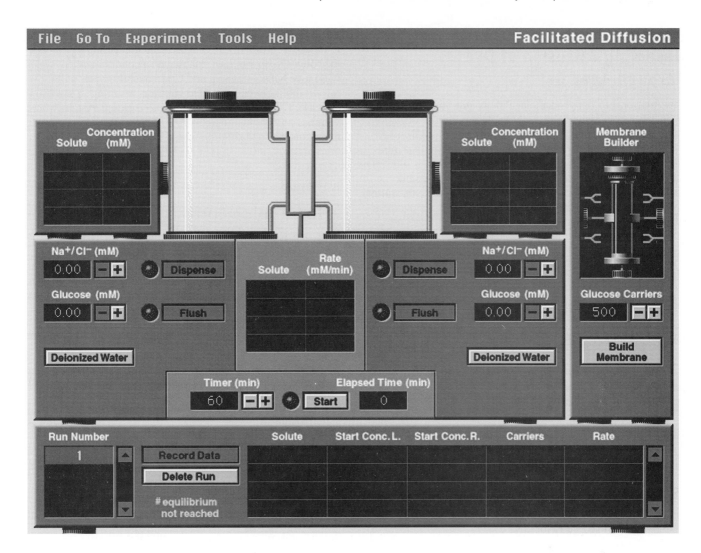

**Figure 5B.2  Opening screen of the Facilitated Diffusion experiment.**

| Chart 2 | Facilitated Diffusion Results | | |
|---|---|---|---|
| | **Number of glucose carrier proteins** | | |
| **Glucose concentration (mg/ml)** | 500 | 700 | 900 |
| 2.00 | | | |
| 8.00 | | | |

What happened to the rate of facilitated diffusion as the number of protein carriers increased? Explain your answer.

What do you think would happen to the transport rate if you put the same concentration of glucose into both beakers instead of deionized water in the right beaker?

Should NaCl have an effect on glucose diffusion? Explain your answer. Use the simulation to arrive at an answer.

## Activity:
# Observing Diffusion of Dye Through Agar Gel

The relationship between molecular weight and the rate of diffusion can be examined easily by observing the diffusion of the molecules of two different types of dye through an agar gel. The dyes used in this experiment are methylene blue, which has a molecular weight of 320 and is deep blue in color, and potassium permanganate, a purple dye with a molecular weight of 158. Although the agar gel appears quite solid, it is primarily (98.5%) water and allows free movement of the diffusing dye molecules through it.

1.  Go to demonstration area 1, and use the millimeter ruler to measure the distance each dye has diffused from each crystal source and record.

Potassium permanganate _____ mm

Methylene blue _____ mm

2.  Make a note of the time the dyes have been diffusing and then compute the rate (mm/min) of each dye's diffusion through the agar gel and record below.

Rate of diffusion:

Potassium permanganate _____ mm/min

Methylene blue _____ mm/min

Which dye diffused more rapidly? _____

What is the relationship between molecular weight and rate of molecular movement (diffusion)?

_____ ■

## Osmosis

A special form of diffusion, the diffusion of water through a semipermeable membrane, is called **osmosis.** Because water can pass through the pores of most membranes, it can move from one side of a membrane to another relatively unimpeded. Osmosis occurs whenever there is a difference in water concentration on the two sides of a membrane.

If we place distilled water on both sides of a membrane, *net* movement of water will not occur; however, water molecules would still move between the two sides of the membrane. In such a situation, we would say that there is no *net* osmosis. The concentration of water in a solution depends on the number of solutes present. Therefore, increasing the solute concentration coincides with a decrease in water concentration. Because water moves down its concentration gradient, it will always move toward the solution with the highest concentration of solutes. Similarly, solutes also move down their concentration gradient. If we position a *fully* permeable membrane (permeable to solutes and water) between two solutions of differing concentrations, then all substances—solutes and water—will diffuse freely, and an equilibrium will

be reached between the two sides of the membrane. However, if we use a semipermeable membrane that is impermeable to the solutes, then we have established a condition where water will move but solutes will not. Consequently, water will move toward the more concentrated solution, resulting in a volume increase. By applying this concept to a closed system where volumes cannot change, we can predict that the pressure in the more concentrated solution would rise.

## Activity:
# Simulating Osmosis

Click the **Experiment** menu and then select **Osmosis.** The opening screen will appear in a few seconds (Figure 5B.3). The most notable difference in this experiment screen concerns meters atop the beakers that measure pressure changes in the beaker they serve. As before, (+) and (−) buttons control solute concentrations in the dispensed solutions.

1.  Drag the 20 MWCO membrane to the holder between the two beakers.

2.  Adjust the NaCl concentration to 8.00 m*M* in the left beaker, and then click the **Dispense** button.

3.  Click **Deionized Water** under the right beaker and then click **Dispense.**

4.  Set the timer to 60 min and then click **Start** to run the experiment. Now click the **Record Data** button to retain your data in the computer's memory and also record the osmotic pressure in Chart 3 below.

5.  Click the membrane to return it to the membrane cabinet.

6.  Repeat steps 1 through 5 with the 50, 100, and 200 MWCO membranes.

Do you see any evidence of pressure changes in either beaker, using any of the four membranes? If so, which ones?

_____

| Chart 3 | Osmosis Results (pressure in mm Hg) | | | |
|---|---|---|---|---|
| | **Membrane (MWCO)** | | | |
| **Solute** | 20 | 50 | 100 | 200 |
| Na⁺Cl⁻ | | | | |
| Albumin | | | | |
| Glucose | | | | |

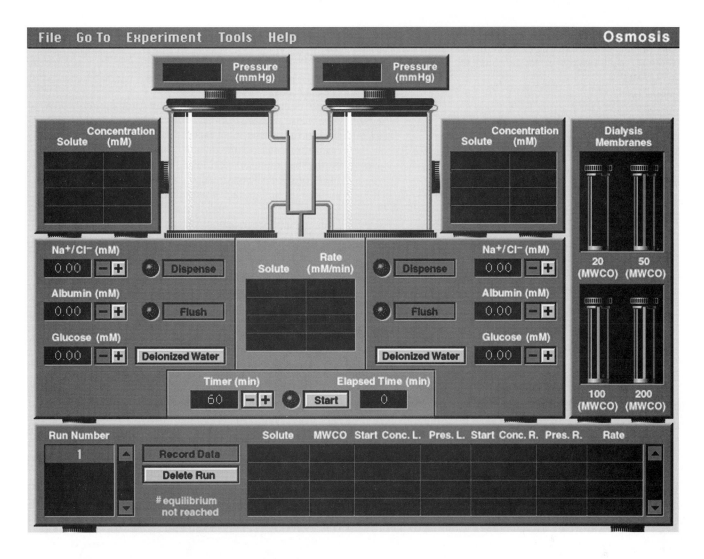

**Figure 5B.3   Opening screen of the Osmosis experiment.**

Does NaCl appear in the right beaker? If so, which membrane(s) allowed it to pass?

_____

_____

7.   Now perform the same experiment for albumin and glucose by repeating steps 1 through 6. In step 2 you will be dispensing 9.00 m$M$ albumin first, and then 10.00 m$M$ glucose, instead of NaCl.

Answer the following questions using the results you recorded in Chart 3. Use the simulation if you need help formulating a response.

Explain the relationship between solute concentration and osmotic pressure.

_____

_____

Will osmotic pressure be generated if solutes are able to diffuse? Explain your answer.

_____

_____

Because the albumin molecule is much too large to pass through a 100 MWCO membrane, you should have noticed the development of osmotic pressure in the left beaker in the albumin run using the 100 MWCO membrane. What do you think would happen to the osmotic pressure if you replaced the deionized water in the right beaker with 9.00 m$M$ albumin in that run? (Both beakers would contain 9.00 m$M$ albumin.)

_____

What would happen if you doubled the albumin concentration in the left beaker using any membrane?

_____

_____

In the albumin run using the 200 MWCO membrane, what would happen to the osmotic pressure if you put 10 m*M* glucose in the right beaker instead of deionized water? Explain your answer.

_____

_____

_____

What if you used the 100 MWCO membrane in the albumin/glucose run described in the previous question?

_____

_____ ■

A c t i v i t y :
## Observing the Osmometer Demonstration

Before continuing on to the filtration simulation, observe the *osmometer demonstration* set up before the laboratory session to follow the movement of water through a membrane (osmosis). Measure the distance the water column has moved during the laboratory period and record it here. The initial position of the molasses meniscus in the thistle tube at the beginning of the laboratory period is marked with wax pencil.

Distance the meniscus has moved: _____ mm. ■

### Filtration

**Filtration** is the process by which water and solutes pass through a membrane from an area of higher hydrostatic (fluid) pressure into an area of lower hydrostatic pressure. Like diffusion, it is a passive process. For example, fluids and solutes filter out of the capillaries in the kidneys into the kid-

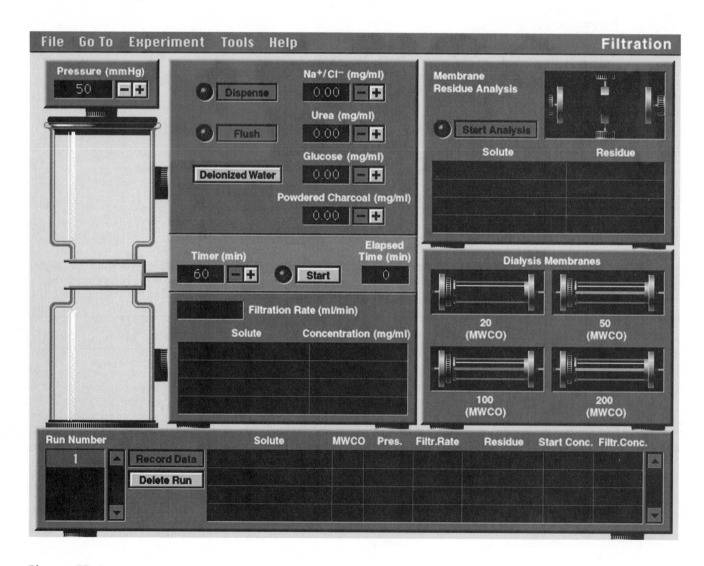

**Figure 5B.4  Opening screen of the Filtration experiment.**

ney tubules because blood pressure in the capillaries is greater than the fluid pressure in the tubules. Filtration is not a selective process. The amount of filtrate—fluids and solutes—formed depends almost entirely on the pressure gradient (the difference in pressure on the two sides of the membrane) and on the size of the membrane pores.

# A c t i v i t y :
## Simulating Filtration

Click the **Experiment** menu and then choose **Filtration.** The opening screen will appear in a few seconds (Figure 5B.4). The basic screen elements resemble the other simulations. The top beaker can be pressurized to force fluid through the filtration membrane into the bottom beaker. Any of the filtration membranes can be positioned in the holder between the beakers by drag-and-drop as in the previous experiments. The solutions you can dispense are listed to the right of the top beaker, and are adjusted by clicking the (+) and (−) buttons. The selected solutions are then delivered to the top beaker by clicking **Dispense.** The top beaker is cleaned and prepared for the next run by clicking **Flush.** You can adjust the timer for any interval between 5 and 300; the elapsed time is shown in the window to the right of the timer. When you click the **Record Data** button, your data is recorded in the computer's memory and is displayed in the data grid at the bottom of the screen.

Solute concentrations in the filtrate are automatically monitored by the *Filtrate Analysis Unit* to the right of the bottom beaker. After a run you can detect the presence of any solute remaining on a membrane by using the *Membrane Residue Analysis* unit located above the membrane cabinet.

1. Click and hold the mouse on the 20 MWCO membrane and drag it to the holder below the top beaker. Release the mouse button to lock the membrane into place.

2. Now adjust the NaCl, urea, glucose, and powdered charcoal windows to 5.00 mg/ml each, and then click **Dispense.**

3. If necessary, adjust the pressure unit atop the beaker until its window reads 50 mm Hg.

4. Set the timer to 60 min and then click **Start.** When the Start button is clicked, the membrane holder below the top beaker retracts, and the solution will flow through the membrane into the beaker below.

5. Watch the Filtrate Analysis Unit for any activity. A rise in detected solute concentration indicates that the solute particles are moving through the filtration membrane. At the end of the run, record the amount of solute present in the *filtrate* (mg/ml) and the filtration *rate* in Chart 4.

6. Now drag the 20 MWCO membrane to the holder in the Membrane Residue Analysis unit. Click **Start Analysis** to begin analysis (and cleaning) of the membrane. Record your results for solute *residue* presence on the membrane (+ for present, − for not present) in Chart 4 and click the **Record Data** button to keep your data in the computer's memory.

7. Click the 20 MWCO membrane again to automatically return it to the membranes cabinet and then click **Flush** to prepare for the next run.

8. Repeat steps 1 through 7 using 50, 100, and 200 MWCO membranes.

| Chart 4 | | Filtration Results (Filtration Rate, Solute Presence or Absence) | | | |
|---|---|---|---|---|---|
| **Membrane (MWCO)** | | | | | |
| **Solute** | | 30 | 40 | 70 | 100 |
| | Rate | | | | |
| NaCl | Filtrate | | | | |
| | Residue | | | | |
| Urea | Filtrate | | | | |
| | Residue | | | | |
| Glucose | Filtrate | | | | |
| | Residue | | | | |
| Powdered charcoal | Filtrate | | | | |
| | Residue | | | | |

Did the membrane's MWCO affect the filtration rate?

_____

Which solute did not appear in the filtrate using any of the membranes?

_____

What would happen if you increased the driving pressure? Use the simulation to arrive at an answer.

_____

Explain how you can increase the filtration rate through living membranes.

_____

_____

By examining the filtration results, we can predict that the

molecular weight of glucose must be greater than _____,

but less than _____. ∎

# Active Transport

Whenever a cell expends cellular energy (ATP) to move substances across its membrane, the process is referred to as an *active transport process.* Substances moved across cell membranes by active means are generally unable to pass by diffusion. There are several possible reasons why substances may not be able to pass through a membrane by diffusion: they may be too large to pass through the pores, they may not be lipid soluble, or they may have to move against rather than with a concentration gradient.

In one type of active transport, substances move across the membrane by combining with a protein carrier molecule; the process resembles an enzyme-substrate interaction. ATP provides the driving force, and in many cases the substances move against concentration or electrochemical gradients or both. Some of the substances that are moved into the cells by such carriers, commonly called **solute pumps,** are amino acids and some sugars. Both solutes are lipid-insoluble and too large to pass through the pores, but are necessary for cell life. On the other hand, sodium ions ($Na^+$) are ejected from the cells by active transport. There is more $Na^+$ outside the cell than inside, so the $Na^+$ tends to remain in the cell unless actively transported out. In the body, the most common type of solute pump is the coupled $Na^+$-$K^+$ pump that moves $Na^+$ and $K^+$ in opposite directions across cellular membranes. 3 $Na^+$ are ejected for every 2 $K^+$ entering the cell.

Engulfment processes such as pinocytosis and phagocytosis also require ATP. In **pinocytosis** (cell drinking), the cell membrane sinks beneath the material to form a small vesicle, which then pinches off into the cell interior. Pinocytosis is most common for taking in liquids containing protein or fat.

In **phagocytosis** (cell eating), parts of the plasma membrane and cytoplasm expand and flow around a relatively large or solid material such as bacteria or cell debris and engulf it, forming a membranous sac called a phagosome. The phagosome is then fused with a lysosome and its contents are digested. In the human body, phagocytic cells are mainly found among the white blood cells and macrophages that act as scavengers and help protect the body from disease-causing microorganisms and cancer cells.

You will examine various factors influencing the function of solute pumps in the following experiment.

### Activity:
## Simulating Active Transport

Click the **Experiment** menu and then choose **Active Transport.** The opening screen will appear in a few seconds (Figure 5B.5). This experiment screen resembles the osmosis experiment screen, except that an ATP dispenser is substituted for the pressure meters atop the beakers. The (+) and (−) buttons control NaCl, KCl, and glucose concentrations in the dispensed solutions. You will use the Membrane Builder to build membranes containing glucose (facilitated diffusion) carrier proteins and active transport $Na^+$-$K^+$ pumps.

In this experiment, we will assume that the left beaker represents the cell's interior and the right beaker represents the extracellular space. The Membrane Builder will insert the $Na^+$-$K^+$ pumps into the membrane so $Na^+$ will be pumped toward the right (out of the cell) while $K^+$ is simultaneously moved to the left (into the cell).

1.   In the Membrane Builder, adjust the number of glucose carriers and the number of $Na^+$-$K^+$ pumps to 500.

2.   Click **Build Membrane,** and then drag the membrane to its position in the membrane holder between the beakers.

3.   Adjust the NaCl concentration to be delivered to the left beaker to 9.00 m*M,* then click the **Dispense** button.

4.   Adjust the KCl concentration to be delivered to the right beaker to 6.00 m*M,* then click **Dispense.**

5.   Adjust the ATP dispenser to 1.00 m*M,* then click **Dispense ATP.** This action delivers the chosen ATP concentration to both sides of the membrane.

6.   Adjust the timer to 60 min, and then click **Start.** Click **Record Data** after each run.

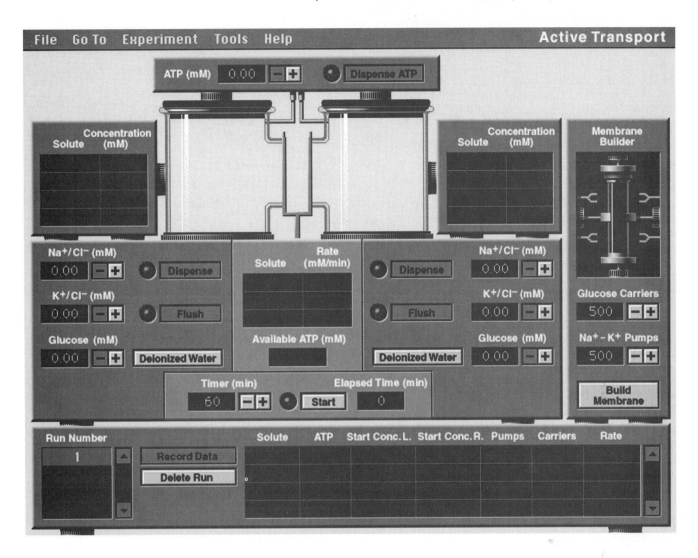

**Figure 5B.5  Opening screen of the Active Transport experiment.**

Watch the solute concentration windows at the side of each beaker for any changes in $Na^+$ and $K^+$ concentrations. The $Na^+$ transport rate slows and then stops before transport has completed. Why do you think that this happens?

_____

What would happen if you did not dispense any ATP?

_____

7.  Click either Flush button to clean both beakers. Repeat steps 3 through 6, adjusting the ATP concentration to 3.00 m$M$ in step 5.

Has the amount of $Na^+$ transported changed?

_____

Do these results support your ideas in step 6 above? _____

What would happen if you decreased the number of $Na^+$-$K^+$ pumps?

_____

Explain how you could show that this phenomenon is not just simple diffusion. (Hint: adjust the $Na^+$ concentration in the right beaker.)

_____

_____

8.  Now repeat steps 1 through 6 dispensing 9.00 m$M$ NaCl into the left beaker and 10.00 m$M$ NaCl into the right beaker (instead of 6.00 m$M$ KCl). Is $Na^+$ transport affected by this change? Explain your answer.

_____

_____

What would happen to the rate of ion transport if we increased the number of $Na^+$-$K^+$ pump proteins?

_____

_____

Would $Na^+$- and $K^+$ transport change if we added glucose solution?

_____

_____

Try adjusting various membrane and solute conditions and attempt to predict the outcome of experimental trials. For example, you could dispense 10 m$M$ glucose into the right beaker instead of deionized water. ■

A c t i v i t y :
## Using Your Data

### Save Your Data

When you have finished with all five experiments, you can save your data set for later review and additional analysis.

1.  Select **Save Data to File** from the **File** menu.

2.  When the dialog box opens, click a drive and/or directory location.

3.  Type a file name into the **File name** box (your instructor may provide you with a name).

4.  Click **Save** to complete the save. If you want to review previously saved data, select **Open Data File** from the **File** menu, and choose your file.

### Print Your Data

If your instructor requires a printed copy of your experiment results, choose **Print Data** from the **File** menu.

1.  Type your name into the space provided.

2.  When you are ready to print, click **OK.** ■

# Skeletal Muscle Physiology: Computer Simulation

exercise

# 16B

## Objectives

1. To define these terms used in describing muscle physiology: *maximal stimulus, treppe, wave summation, multiple motor unit summation, tetanus*

2. To identify two ways that the mode of stimulation can affect muscle force production.

3. To plot a graph relating stimulus strength and twitch force to illustrate graded muscle response.

4. To explain how slow, smooth, sustained contraction is possible in a skeletal muscle.

5. To graphically understand the relationships between passive, active, and total forces.

6. To identify the conditions under which muscle contraction is isometric or isotonic.

7. To describe in terms of length and force the transitions between isometric and isotonic conditions during a single muscle twitch.

8. To describe the effects of resistance and starting length on the initial velocity of shortening.

9. To explain why muscle force remains constant during isotonic shortening.

10. To explain experimental results in terms of muscle structure.

## Materials

Minimum equipment required:

**Windows:**

❑ 486/66 MHz or better recommended
❑ Windows 95/98 recommended
❑ SVGA display (256 colors at 640 X 480)
❑ Minimum 16 MB available RAM
❑ Double-speed CD-ROM drive (quad-speed or higher recommended)
❑ Sound card
❑ Speakers or headphones
❑ Printer

**Macintosh:**

❑ 68040 processor or Power Macintosh recommended

❑ System 7.1 or higher
❑ Minimum 16 MB available RAM
❑ 13-inch or larger color monitor (640 X 480 resolution)
❑ Double-speed CD-ROM drive (quad-speed or higher recommended)
❑ Printer

**Software:**

❑ Benjamin/Cummings PhysioEx CD-ROM—Skeletal Muscle Physiology module

Skeletal muscles are composed of hundreds to thousands of individual cells, each doing their share of work in the production of force. As their name suggests, skeletal muscles move the skeleton. Skeletal muscles are remarkable machines; while allowing us the manual dexterity to create magnificent works of art, they are also capable of generating the brute force needed to lift a 100-lb sack of concrete. When a skeletal muscle from an experimental animal is electrically stimulated, it behaves in the same way as a stimulated muscle in the intact body, that is, *in vivo*. Hence, such an experiment gives us valuable insight into muscle behavior.

This set of computer simulations demonstrates many important physiological concepts of skeletal muscle contraction. The program graphically provides all the equipment and materials necessary for you, the investigator, to set up experimental conditions and observe the results. In student-conducted laboratory investigations there are many ways to approach a problem, and the same is true of these simulations. The instructions will guide you in your investigation, but you should also try out alternate approaches to gain insight into the logical methods used in scientific experimentation.

Try this approach: As you work through the simulations for the first time, follow the instructions closely and answer the questions posed as you go. Then try asking "What if . . . ?" questions to test the validity of your theories. The major advantages of these computer simulations are that the muscle cannot be accidentally damaged, lab equipment will not break down at the worst possible time, and you will have ample time to think critically about the processes being investigated.

Because you will be working with a simulated muscle and oscilloscope display, you need to watch both carefully during the experiments. Think about what is happening in each situation. You need to understand how you are experimentally manipulating the muscle in order to understand your results.

**13**

# Getting Started

Begin by making sure you have the computer equipment and software listed in the Materials section on p. 13.

1.   Quit all applications currently running on your computer.

2.   Insert the PhysioEx CD-ROM into the CD-ROM drive and keep PhysioEx in the drive the entire time you use the program.

3.   Follow the instructions below for your computer type.

**Windows 95/Windows 98**
The program will launch automatically after you load the CD-ROM into the drive.
If autorun is disabled on your machine follow these instructions:
1.   Double-click the My Computer icon on the Windows desktop.
2.   Double-click the PhysioEx CD icon.
3.   Look for the icon that looks like a tiny movie projector and double-click it.

**Macintosh**
Double-click the PhysioEx CD icon that appears in the window on your desktop to launch the program.

# Electrical Stimulation

A contracting skeletal muscle will produce force and/or shortening when nervous or electrical stimulation is applied. Unlike single cells or motor units, which follow the all-or-none law of muscle physiology, a whole muscle responds to stimuli with a graded response. A motor unit consists of a motor neuron and all the muscle cells it innervates. Hence, activation of the neuron innervating a single motor unit will cause all muscle cells in that unit to fire simultaneously in an all-or-none fashion. The graded contractile response of a whole muscle reflects the number of motor units firing at a given time.  Strong muscle contraction implies that many motor units are activated and each unit has maximally contracted. Weak contraction means that few motor units are active; however, the activated units are maximally contracted. By increasing the number of motor units firing, we can produce a steady increase in muscle force, a process called recruitment or motor unit summation.

Regardless of the number of motor units activated, a single contraction of skeletal muscle is called a muscle twitch. A tracing of a muscle twitch is divided into three phases: latent, contraction, and relaxation. The latent phase is a short period between the time of stimulation and the beginning of contraction. Although no force is generated during this interval, chemical changes occur intracellularly in preparation for contraction. During contraction, the myofilaments are sliding past each other and the muscle shortens. Relaxation takes place when contraction has ended and the muscle returns to its normal resting state and length.

The first activity you will conduct simulates an isometric, or fixed length, contraction of an isolated skeletal muscle. This activity allows you to investigate how the strength and frequency of an electrical stimulus affect whole muscle function. Note that these simulations involve indirect stimulation by an electrode placed on the surface of the muscle. This differs from the situation *in vivo* where each fiber in the muscle receives direct stimulation via a nerve ending. In other words, increasing the intensity of the electrical stimulation mimics how the nervous system increases the number of motor units activated.

## Single Stimulus

Choose **Skeletal Muscle Physiology** from the main menu. The opening screen will appear in a few seconds (Figure 16B.1). The oscilloscope display, the grid at the top of the screen, is the most important part of the screen because it graphically displays the contraction data for analysis. Time is displayed on the horizontal axis. A full sweep is initially set at 200 msec. However, you can adjust the sweep time from 200 msec to 1000 msec by clicking and dragging the **200** msec button at the lower right corner of the oscilloscope display to the left to a new position on the time axis. The force (in grams) produced by muscle contraction is displayed on the vertical axis. Clicking the **Clear Tracings** button erases all muscle twitch tracings from the oscilloscope display.

The *electrical stimulator* is the equipment seen just beneath the oscilloscope display. Clicking **Stimulate** delivers the electrical shock to the muscle through the electrodes lying on the surface of the muscle. Stimulus voltage is set by clicking the (+) or (−) buttons next to the voltage window. Three small windows to the right of the Stimulate button display the force measurements. *Active force* is produced during muscle contraction, while *passive force* results from the muscle being stretched (much like a rubber band). The *total force* is the sum of active and passive forces. The red arrow at the left of the oscilloscope display is an indicator of passive force. After the muscle is stimulated, the **Measure** button at the right edge of the electrical stimulator becomes active. When the Measure button is clicked, a vertical orange line will be displayed at the left edge of the oscilloscope display. Clicking the arrow buttons below the Measure button moves the orange line horizontally across the screen. The Time window displays the difference in time between the zero point on the X-axis and the intersection between the orange measure line and the muscle twitch tracing.

The muscle is suspended in the support stand to the left of the oscilloscope display. The hook through the upper tendon of the muscle is part of the force transducer, which measures the force produced by the muscle. The hook through the lower tendon secures the muscle in place. The weight cabinet just below the muscle support stand is not active in this experiment; it contains weights you will use in the isotonic contraction part of the simulation. You can adjust the starting length of the muscle by clicking the (+) or (−) buttons located next to the Muscle Length display window.

When you click the **Record Data** button in the data collection unit below the electrical stimulator, your data is recorded in the computer's memory and is displayed in the

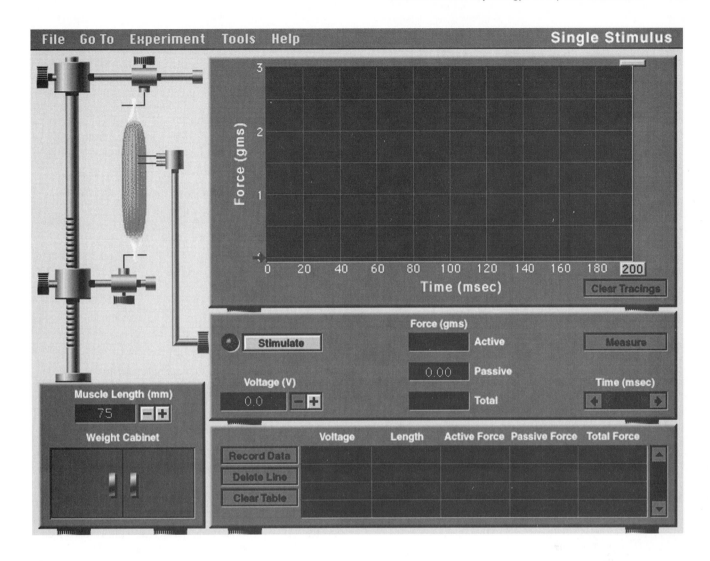

**Figure 16B.1  Opening screen of the Single Stimulus experiment.**

data grid at the bottom of the screen. Data displayed in the data grid include the voltage, muscle length, and active, passive, and total force measurements. If you are not satisfied with a single run, you can click **Delete Line** to erase a single line of data. Clicking the **Clear Table** button will remove all accumulated data in the experiment and allow you to start over.

### Activity:
## Practicing Generating a Tracing

1.   Click the **Stimulate** button once. Because the beginning voltage is set to zero, no muscle activity should result. You will see a blue line moving across the bottom of the oscilloscope display. This blue line will indicate muscle force in the experiments. If you are using Windows 3.1 or an older Macintosh and the tracings move too slowly across the screen, click and hold the **200** button at the lower right corner of the oscilloscope and drag it to the left to the 40 msec mark and

release it. This action resets the total sweep time to 1000 msec to speed up the display time.

2.   Click and hold the (+) button beneath the Stimulate button until the voltage window reads 3.0 volts. Click **Stimulate** once. You will see the muscle react, and a contraction tracing will appear on the screen. Notice that the muscle tracing color alternates between blue and yellow each time the Stimulate button is clicked to enhance the visual difference between twitch tracings. You can click the **Clear Tracings** button as needed to clean up the oscilloscope display. To retain your data, click the **Record Data** button at the end of each stimulus.

3.   Change the voltage to 5.0 volts and click **Stimulate** again. Notice how the force of contraction also changes. Identify the latent, contraction, and relaxation phases in the tracings.

Feel free to experiment with anything that comes to mind to get a sense of how whole muscle responds to an electrical stimulus. ■

A c t i v i t y :
# Determining the Latent Period

1. Click **Clear Tracings** to erase the oscilloscope display. The voltage should be set to 5.0 volts.

2. Drag the **200** msec button to the right edge of the oscilloscope.

3. Click the **Stimulate** button once and allow the tracing to complete (this may be quite slow on Windows 3.1 or an older Macintosh).

4. When the tracing is completed, click the **Measure** button. Click the right arrow button next to the Time window repeatedly until you notice the first increase in the active force window. Click the left arrow button next to the Time window until the active force window reads zero. The time now displayed in the Time window represents the difference between when the muscle was stimulated and the beginning of force production. This time represents the latent period of muscle contraction.

How long is the latent period? _____ msec

What occurs in the muscle during this apparent lack of activity?

_____

_____ ▪

A c t i v i t y :
# Determining the Threshold (Minimal) Stimulus

1. Click **Clear Tracings** to erase the oscilloscope display.

2. Set the voltage control to 0.0 volts by clicking the (−) button next to the voltage window.

3. Increase the voltage by 0.1 volt and click **Stimulate.**

4. Repeat step 3 until you see a number greater than 0.0 appear in the Active Force window. This number indicates that a contraction has occurred. The number in the voltage window represents the threshold voltage, below which no contraction occurs.

What is the threshold voltage? _____ V ▪

A c t i v i t y :
# Investigating Graded Muscle Response to Increased Stimulus Intensity

1. Examine the tracings just recorded in the previous activity.

2. Now stimulate the muscle several times using increasingly higher voltages. Start at 6.0 volts and increase the voltage in approximately 1.0-volt increments. If the display is slow, drag the **200** msec button back to the 40 msec mark.

Notice that as voltage is increased, the amount of force produced by the muscle also increases. As more voltage is delivered to the whole muscle, greater numbers of muscle fibers are activated, thereby increasing the total force produced by the muscle. This result mimics the muscle activity occurring *in vivo* where the recruitment of additional motor units increases the total force produced. This phenomenon is called *multiple motor unit summation.*

Is there a stimulus voltage above which there appears to be no further increase in muscle contraction?

_____

Why is this so?

_____

_____

3. Click **Clear Tracings** to prepare for the next experiment. You will perform a systematic study of force produced over the entire range of stimulus voltages in 0.5-volt increments up to 10.0 volts.

4. Set the voltage to 0.0 and click **Stimulate.**

5. Click **Record Data**. If you decide to redo a single stimulus, choose the data line in the grid and click **Delete Line** to erase that single line of data. If you want to repeat the entire experiment, click the **Clear Table** button to erase all data recorded to that point.

6. Repeat steps 4 and 5, increasing the voltage by 0.5 each time until you reach the maximum voltage of 10.0.

Based on the twitch tracings you see in the display, is a muscle twitch an all-or-none response?

_____

Explain your answer.

_____

_____

_____

_____

7. Observe the twitch tracings. Click the **Tools** menu and then choose **Plot Data.** Note that the oscilloscope tracings will be automatically erased from the oscilloscope display, but your data will be retained in the grid.

8. Use the slider bars to display Active Force on the Y-axis and Voltage on the X-axis. Click **Draw It** to graph your collected data. Estimate the maximal stimulus by looking at your graph.

What is the maximal stimulus? _____ V

This voltage represents the lowest stimulus intensity or strength necessary to activate all cells within the muscle and is called the **maximal stimulus.** The maximum contraction thus produced is called the maximal response.

9. When finished, click the close box at the top of the plot window. ■

## Multiple Stimulus

Choose **Multiple Stimulus** from the **Experiment** menu. The opening screen will appear in a few seconds (Figure 16B.2).

The only significant change to the on-screen equipment is found in the electrical stimulator. The measuring equipment has been removed and other controls have been added: The **Multiple Stimulus** button is a toggle that allows you to

alternately start and stop the electrical stimulator. When Multiple Stimulus is first clicked, its name changes to Stop Stimulus and electrical stimuli are delivered to the muscle at the rate specified in the Stimuli/sec window until the muscle completely fatigues or the stimulator is turned off. The stimulator is turned off by clicking the **Stop Stimulus** button. The stimulus rate is adjusted by clicking the (+) or (−) buttons next to the stimuli/sec window.

## Activity:
## Investigating Treppe

When a muscle first contracts, the force it is able to produce is less than the force it is able to produce in subsequent contractions within a relatively narrow time span. A myogram, a recording of a muscle twitch, reveals this phenomenon as the **treppe,** or staircase, effect. For the first few twitches, each

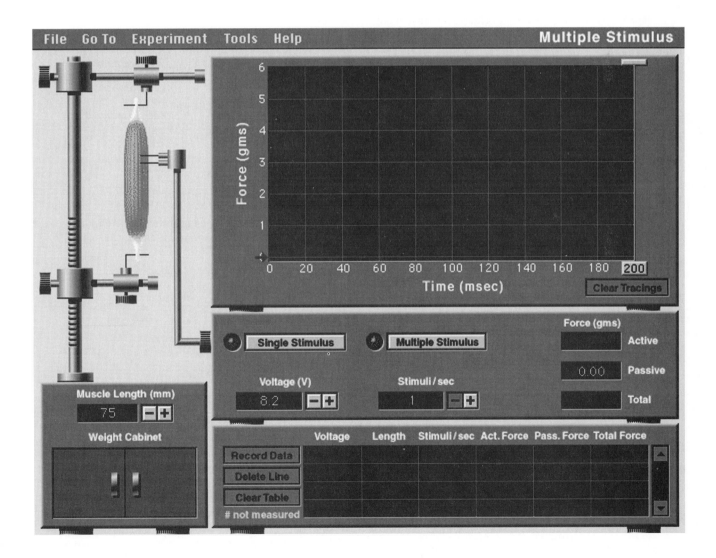

**Figure 16B.2 Opening screen of the Multiple Stimulus experiment.**

successive stimulation produces slightly more force than the previous contraction as long as the muscle is allowed to fully relax between stimuli, and the stimuli are delivered relatively close together. Treppe is thought to be caused by increased efficiency of the enzyme systems within the cell and increased availability of intracellular calcium.

1. The voltage should be set to 8.2 volts and the muscle length should be 75 mm.

2. Drag the **200** msec button to the center of the X-axis time range.

3. Be sure that you fully understand the following three steps before you proceed.

• Click **Single Stimulus.** Watch the twitch tracing carefully.

• After the tracing shows that the muscle has completely relaxed, immediately click **Single Stimulus** again.

• When the second twitch completes, click **Single Stimulus** once more and allow the tracing to complete.

What happens to force production with each subsequent stimulus?

_____

_____ ■

### Activity:
# Investigating Wave Summation

As demonstrated in the single-stimulus activity earlier, multiple motor unit summation is one way to increase the amount of force produced by muscle. **Multiple motor unit summation** relies on increased stimulus *intensity* in this simulation. Another way to increase force is by wave, or temporal, summation. **Wave summation** is achieved by increasing the stimulus *frequency,* or rate of stimulus delivery to the muscle. Wave summation occurs because the muscle is already in a partially contracted state when subsequent stimuli are delivered.

**Tetanus** can be considered an extreme form of wave summation that results in a steady, sustained contraction. In effect, the muscle does not have any chance to relax because it is being stimulated at such a high frequency. This "fuses" the force peaks so that we observe a smooth tracing.

1. Click **Clear Tracings** to erase the oscilloscope display.

2. Set and keep the voltage at the maximal stimulus (8.2 volts) and the muscle length at 75 mm.

3. Drag the **200** msec button to the right edge of the oscilloscope display unless you are using a slower computer.

4. Click **Single Stimulus,** and then click **Single Stimulus** again when the muscle has relaxed about halfway.

Is the peak force produced in the second contraction greater than that produced by the first stimulus?

_____

5. Try stimulating again at greater frequencies by clicking the Single Stimulus button several times in rapid succession.

Is the total force production even greater? _____

6. To see if you can produce smooth, sustained contraction at Active Force = 2 gms, try rapidly clicking **Single Stimulus** several times.

Is it possible to produce a smooth contraction (fused force peaks), or does the force rise and fall periodically?

_____

7. Using the same method as in step 6, try to produce a smooth contraction at 3 gms.

Is the trace smoother (less height between peaks and valleys) this time?

_____

Because there is a limit to how fast you can manually click the Single Stimulus button, what do you think would happen to the smoothness of the tracing if you could click even faster?

_____

_____

_____

8. So far you have been using the maximal stimulus. What do you think would happen if you used a lower voltage?

_____

_____

_____

_____

9. Use the concepts of motor unit summation and stimulus frequency to explain how human skeletal muscles work to achieve smooth, steady contractions at all desired levels of force.

_____

_____

_____

_____ ■

# Investigating Fusion Frequency

1. Click **Clear Tracings** to erase the oscilloscope display.

2. The voltage should be set to 8.2 volts and the muscle length should be 75 mm.

3. Adjust the stimulus rate to 30 stimuli/sec.

4. The following steps constitute a single "run." Become familiar with the procedure for completing a run before continuing.

- Click **Multiple Stimulus.**

- When the tracing is close to the right side of the screen, click **Stop Stimulus** to turn off the stimulator.

- Click **Record Data** to retain your data in the grid in the bottom of the screen and the computer's memory.

If you decide to redo a single run, choose the data line in the grid and click **Delete Line** to erase that single line of data. If you want to repeat the entire experiment, click the **Clear Table** button to erase all data recorded thus far.

Describe the appearance of the tracing.

_____

_____

5. Repeat step 4, increasing the stimulation rate by 10 stimuli/sec each time up to 150 stimuli/sec.

How do the tracings change as the stimulus rate is increased?

_____

6. When you have finished observing the twitch tracings, click the **Tools** menu and then choose **Plot Data.** Notice that the oscilloscope tracings will be automatically erased from the oscilloscope display, but your data will be retained in the grid.

7. Set the Y-axis slider to display Active Force and the X-axis slider to display Stimuli/sec. Click **Draw It** to display a graph of your collected data.

From your graph, estimate the stimulus rate above which there appears to be no significant increase in force.

_____ stimuli/sec

This rate is the fusion frequency, also called tetanus.

8. When finished, click the close box at the top of the plot window.

9. Reset the stimulus rate to the fusion frequency.

10. Try to produce a smooth contraction at Force = 2 gms and Force = 3 gms by adjusting only the stimulus intensity, or voltage, using the following procedure.

- Decrease the voltage to a starting point of 1.0 volt and click **Multiple Stimulus.**

- Click **Stop Stimulus** to turn off the stimulator when the tracing is near the right side of the oscilloscope display.

- If the force produced is not smooth and continuous at the desired level of force, increase the voltage in 0.1-volt increments and stimulate as above until you achieve a smooth force at 2 gms and again at 3 gms.

What stimulus intensity produced smooth force at Force = 2 gms?

_____ V

Which intensity produced smooth contraction at Force = 3 gms?

_____ V

Explain what must be happening in the muscle to achieve smooth contraction at different force levels.

_____

_____

_____

_____

_____ ■

# Investigating Muscle Fatigue

A prolonged period of sustained contraction will result in muscle fatigue, a condition in which the tissue has lost its ability to contract. Fatigue results when a muscle cell's ATP consumption is faster than its production. Consequently, increasingly fewer ATP molecules are available for the contractile parts within the muscle cell.

If you are using Windows 3.1 or an older Macintosh and the tracings move too slowly across the screen, click and drag the **200** button at the lower right corner of the oscilloscope to the 40 msec mark to speed up the display time.

1. Click **Clear Tracings** to erase the oscilloscope display.

2. The voltage should be set to 8.2 volts and the muscle length should be 75 mm.

3. Adjust the stimulus rate to 120 stimuli/sec.

4. Click **Multiple Stimulus** and allow the tracing to sweep through three screens and then click **Stop Stimulus** to stop the stimulator.

Why does the force begin to fall with time? Note that a fall in force indicates muscle fatigue.

_____

_____

5. Click **Clear Tracings** to erase the oscilloscope display. Keep the same settings as before.

6. You will be clicking **Multiple Stimulus** on and off three times to demonstrate fatigue with recovery. Read the steps below before proceeding.

• Click **Multiple Stimulus.**

• When the tracing reaches the middle of the screen, briefly turn off the stimulator by clicking **Stop Stimulus,** then immediately click **Multiple Stimulus** again.

• You will see a dip in the force tracing where you turned the stimulator off and then on again. The force tracing will continue to drop as the muscle fatigues.

• Before the muscle fatigues completely, repeat the on/off cycle twice more without clearing the screen.

Turning the stimulator off allows a small measure of recovery. The muscle will produce force for a longer period if the stimulator is briefly turned off than if the stimulations were allowed to continue without interruption. Explain why.

_____

_____

_____

7. To see the difference between continuous multiple stimulation and multiple stimulation with recovery, click **Multiple Stimulus** and let the tracing fall without interruption to zero force. This tracing will follow the original myogram exactly until the first "dip" is encountered, after which you will notice a difference in the amount of force produced between the two runs.

Describe the difference between the current tracing and the myogram generated in step 6.

_____

_____

_____

_____

_____

_____

_____ ▪

# Isometric Contraction

Isometric contraction is the condition in which muscle length does not change regardless of the amount of force generated by the muscle (*iso* = same, *metric* = length). This is accomplished experimentally by keeping both ends of the muscle in a fixed position while stimulating it electrically. Resting length (length of the muscle before contraction) is an important factor in determining the amount of force that a muscle can develop. Passive force is generated by stretching the muscle, and is due to the elastic properties of the tissue itself. Active force is generated by the physiological contraction of the muscle. Think of the muscle as having two force properties: it exerts passive force when it is stretched (like a rubber band exerts passive force), and active force when it contracts. Total force is the sum of passive and active forces, and it is what we experimentally measure.

This simulation allows you to set the resting length of the experimental muscle and stimulate it with individual maximal stimulus shocks. A graph relating the forces generated to the length of the muscle will be automatically plotted as you stimulate the muscle. The results of this simulation can then be applied to human muscles in order to understand how optimum resting length will result in maximum force production. In order to understand why muscle tissue behaves as it does, it is necessary to comprehend contraction at the cellular level. Hint: If you have difficulty understanding the results of this exercise, review the sliding filament model of muscle contraction. Then think in terms of sarcomeres that are too short, too long, or just the right length.

Choose **Isometric Contraction** from the **Experiment** menu. The opening screen will appear in a few seconds (Figure 16B.3). Notice that the oscilloscope is now divided into two parts. The left side of the scope displays the muscle twitch tracing. The active, passive, and total force data points are plotted on the right side of the screen.

### Activity:
## Investigating Isometric Contraction

1. The voltage should be set to the maximal stimulus (8.2 volts) and the muscle length should be 75 mm.

2. To see how the equipment works, stimulate once by clicking **Stimulate.** You should see a single muscle twitch tracing on the left oscilloscope display, and three data points representing active, passive, and total force on the right display. The yellow box represents the total force and the red dot it contains symbolizes the superimposed active force. The green square represents the passive force data point.

3. Try adjusting the muscle length by clicking the (+) or (−) buttons located next to the Muscle Length window and watch the effect on the muscle.

4. When you feel comfortable with the equipment, click **Clear Tracings** and **Clear Plot.**

5. Now stimulate at different muscle lengths using the following procedure.

• Shorten the muscle to a length of 50 mm by clicking the (−) button next to the Muscle Length window.

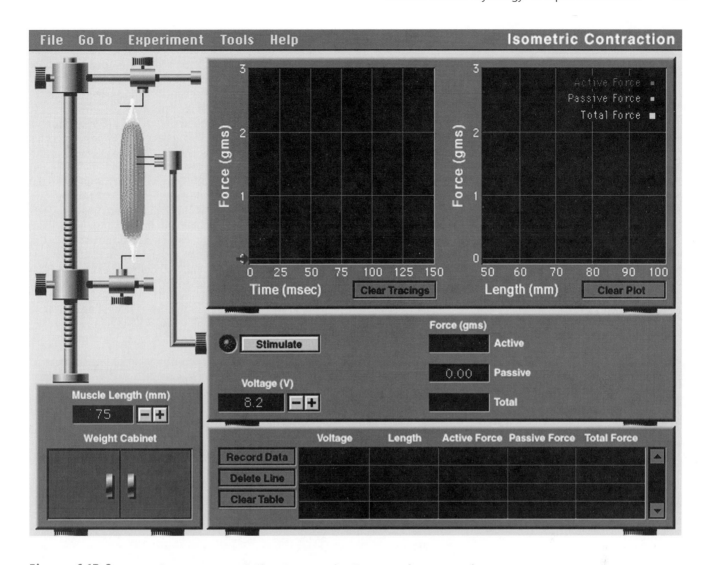

**Figure 16B.3  Opening screen of the Isometric Contraction experiment.**

- Click **Stimulate** and when the tracing is complete, click **Record Data.**

- Repeat the **Stimulate** and **Record Data** sequence, increasing the muscle length by 2 mm each time until you reach the maximum muscle length of 100 mm.

6.  Carefully examine the active, passive, and total force plots in the right oscilloscope display.

What happens to the passive and active forces as the muscle length is increased from 50 mm to 100 mm?

Passive force:

_____

_____

Active force:

_____

_____

Total force:

_____

_____

Explain the dip in the total force curve. (Hint: keep in mind you are measuring the sum of active and passive forces.)

_____

_____

_____

_____

_____ ∎

# Isotonic Contraction

During isotonic contraction, muscle length changes, but the force produced stays the same (*iso* = same, *tonic* = force). Unlike the isometric exercise in which both ends of the muscle are held in a fixed position, one end of the muscle remains free in the isotonic contraction exercise. Different weights can then be attached to the free end while the other end is fixed in position on the force transducer. If the weight is not too great, the muscle will be able to lift it with a certain velocity. You can think of lifting an object from the floor as an example: if the object is light it can be lifted quickly (high velocity), whereas a heavier weight will be lifted with a slower velocity. Try to transfer the idea of what is happening in the simulation to the muscles of your arm when you lift a weight. The two important variables in this exercise are starting length of the muscle and resistance (weight) applied. You have already examined the effect of starting length on muscle force production in the previous exercise. Now you will change both muscle length and resistance to investigate how such changes affect the speed of skeletal muscle shortening. Both variables can be independently altered and the results are graphically presented on the screen.

Choose **Isotonic Contraction** from the **Experiment** menu. The opening screen will appear in a few seconds (Figure 16B.4). The general operation of the equipment is the same as in the previous experiments. In this simulation, the weight cabinet doors are open. You will attach weights to the lower tendon of the muscle by clicking and holding the mouse on any weight in the cabinet and then dragging-and-dropping the weight's hook onto the lower tendon. The length window displays the muscle length achieved when the muscle is stretched by hanging a weight from it's lower tendon. You can click the (+) and (−) buttons next to the Platform Height window to change the position of the platform

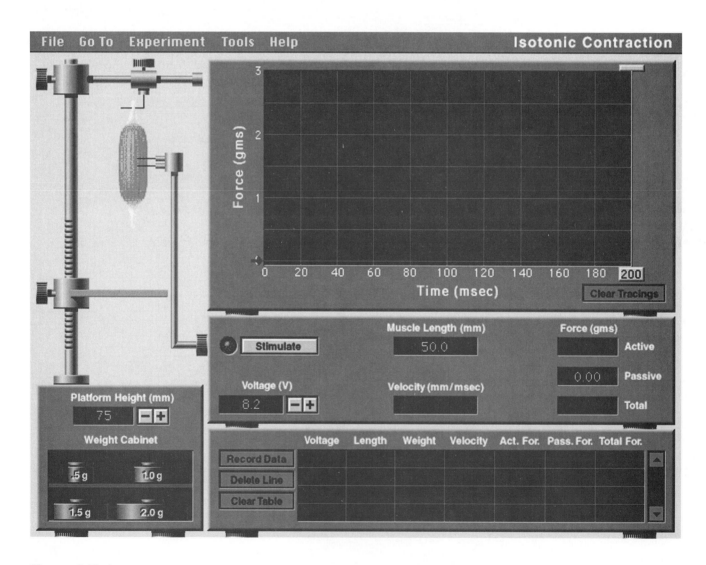

**Figure 16B.4** **Opening screen of the Isotonic Contraction experiment.**

on which the weight rests. Click on the weight again to automatically return it to the weight cabinet. The electrical stimulator displays the initial velocity of muscle shortening in the small window to the right of the Voltage control.

If you are using Windows 3.1 or an older Macintosh and the tracings move too slowly across the screen, click and hold the **200** button at the lower right corner of the oscilloscope and drag it to the left to the 40 msec mark and release it. This action resets the total sweep time to 1000 msec to speed up the display time.

### Activity:
## Investigating the Effect of Load on Skeletal Muscle

1.  Set the voltage to the maximal stimulus (8.2 volts).

2.  Drag-and-drop the .5-g weight onto the muscle's lower tendon.

3.  Platform height should be 75 mm.

4.  Click **Stimulate** and simultaneously watch the muscle action and the oscilloscope tracing.

5.  Click the **Record Data** button to retain and display the data in the grid.

What do you see happening to the muscle during the flat part of the tracing? Click **Stimulate** to repeat if you wish to see the muscle action again.

_____

_____

Does the force the muscle produces change during the flat part of the tracing (increases, decreases, or stays the same)?

_____

Describe the muscle activity during the flat part of the tracing in terms of isotonic contraction and relaxation.

_____

_____

_____

_____

Circle the correct terms in parentheses in the following sequence: The force rises during the first part of the muscle tracing due to (isometric, isotonic) contraction while the fall in force on the right side of the tracing corresponds to (isometric, isotonic) relaxation.

6.  Return the .5-g weight to the cabinet. Drag the 1.5-g weight to the muscle. Click **Stimulate** and then click **Record Data.**

Which of the two weights used so far results in the highest initial velocity of shortening?

_____

7.  Repeat step 6 for the remaining two weights.

8.  Choose **Plot Data** from the **Tools** menu.

9.  Set Weight as the X-axis and Force Produced as the Y-axis by dragging the slider bars. Click **Draw It.**

What does the plot reveal about the resistance and the initial velocity of shortening?

_____

_____

_____

10. Close the plot window and then click **Clear Table** in the data control unit at the bottom of the screen. Click **Yes** when you are asked if you want to erase all data in the table.

11. Return the current weight to the weight cabinet.

12. Attach the 1.5-g weight to the muscle and run through the range of starting lengths from 60–90 mm in 5-mm increments. Be sure to click **Record Data** after each stimulus.

13. After all runs have been completed, choose **Plot Data** from the **Tools** menu.

14. Set the Length as the X-axis and the Velocity as the Y-axis by dragging the slider bars.

Describe the relationship between starting length and initial velocity of shortening.

_____

_____

_____

_____

Do these results support your ideas in the isometric contraction part of this exercise?

_____

15. Close the plot window.

Can you set up a contraction that is completely isometric?

_____ One that is completely isotonic? _____

Explain your answers.

———————————————————

———————————————————

———————————————————

——————————————————— ▪

## Activity:
# Using Your Data

## Save Your Data

When you have finished with all experiments, you can save your data set for later review and additional analysis.

1. Select **Save Data to File** from the **File** menu.

2. When the dialog box opens, click a drive and/or directory location.

3. Type a file name into the File name box (your instructor may provide you with a name).

4. Click **Save** to complete the save. If you want to review previously saved data, select **Open Data File** from the **File** menu, and choose your file.

## Print Your Data

If your instructor requires a hard copy of your experiment results, choose **Print Data** from the **File** menu.

1. Type your name into the provided space.

2. When you are ready to print, click **OK**. ▪

# Cardiovascular Dynamics: Computer Simulation

The physiology of human blood circulation can be divided into two distinct but remarkably harmonized processes: (1) the pumping of blood by the heart, and (2) the transport of blood to all body tissues via the vasculature, or blood vessels. Blood supplies all body tissues with the substances needed for survival, so it is vital that blood delivery is ample for tissue demands.

## The Mechanics of Circulation

To understand how blood is transported throughout the body, let's examine three important factors influencing how blood circulates through the cardiovascular system: blood flow, blood pressure, and peripheral resistance.

**Blood flow** is the amount of blood moving through a body area or the entire cardiovascular system in a given amount of time. While total blood flow is determined by cardiac output (the amount of blood the heart is able to pump per minute), blood flow to specific body areas can vary dramatically in a given time period. Organs differ in their requirements from moment to moment, and blood vessels constrict or dilate to regulate local blood flow to various areas in response to the tissue's immediate needs. Consequently, blood flow can increase to some regions and decrease to other areas at the same time.

**Blood pressure** is the force blood exerts against the wall of a blood vessel. Owing to cardiac activity, pressure is highest at the heart end of any artery. Because of the effect of peripheral resistance, which will be discussed shortly, pressure within the arteries (or any blood vessel) drops as the distance (vessel length) from the heart increases. This pressure gradient causes blood to move from and then back to the heart, always moving from high- to low-pressure areas.

**Peripheral resistance** is the opposition to blood flow resulting from the friction developed as blood streams through blood vessels. Three factors affect vessel resistance: blood viscosity, vessel radius, and vessel length.

**Blood viscosity** is a measure of the "thickness" of the blood, and is caused by the presence of proteins and formed elements in the plasma (the fluid part of the blood). As the viscosity of a fluid increases, its flow rate through a tube decreases. Blood viscosity in healthy persons normally does not change, but certain conditions such as too many or too few blood cells may modify it.

## Objectives

1. To define the following: blood flow; viscosity; peripheral resistance; systole; diastole; end diastolic volume; end systolic volume; stroke volume; cardiac output.
2. To explore cardiovascular dynamics using an experimental setup to simulate a human body function.
3. To understand that heart and blood vessel functions are highly coordinated.
4. To comprehend that pressure differences provide the driving force that moves blood through the blood vessels.
5. To recognize that body tissues may differ in their blood demands at a given time.
6. To identify the most important factors in control of blood flow.
7. To comprehend that changing blood vessel diameter can alter the pumping ability of the heart.
8. To examine the effect of stroke volume on blood flow.

## Materials

Minimum computer hardware required:

**Windows:**

- ❏ 486/66MHz or better recommended
- ❏ Windows 95/98 recommended
- ❏ SVGA display (256 colors at 640 × 480)
- ❏ Minimum 16 MB available RAM
- ❏ Double-speed CD-ROM drive (quad speed or higher recommended)
- ❏ Sound card
- ❏ Speakers or headphones
- ❏ Printer

**Macintosh:**

- ❏ 68040 processor or Power Macintosh recommended
- ❏ System 7.1 or higher
- ❏ Minimum 16 MB available RAM

- ❑ 13-inch or larger color monitor (640 × 480 resolution)
- ❑ Double-speed CD-ROM drive (quad speed or higher recommended)
- ❑ Printer
- ❑ **Software:** Benjamin/Cummings *PhysioEx* CD-ROM—Cardiovascular Dynamics module

Controlling *blood vessel radius* (one-half of the diameter) is the principal method of blood flow control. This is accomplished by contracting or relaxing the smooth muscle within the blood vessel walls. To see why radius has such a pronounced effect on blood flow, we need to explore the physical relationship between blood and the vessel wall. Blood in direct contact with the vessel wall flows relatively slowly because of the friction, or drag, between the blood and the lining of the vessel. In contrast, fluid in the center of the vessel flows more freely because it is not "rubbing" against the vessel wall. When we contrast large- and small-radius vessels, we see that proportionately more blood is in contact with the wall of small vessels, hence blood flow is notably impeded in small-radius vessels.

Although *vessel length* does not ordinarily change in a healthy person, any increase in vessel length causes a corresponding flow decrease. This effect is principally caused by friction between blood and the vessel wall. Consequently, given two blood vessels of the same diameter, the longer vessel will have more resistance, and thus a reduced blood flow.

### The Effect of Blood Pressure and Vessel Resistance on Blood Flow

Poiseuille's equation describes the relationship between pressure, vessel radius, viscosity, and vessel length on blood flow:

$$\text{Blood flow } (\Delta Q) = \frac{\pi \Delta P \, r^4}{8 \eta l}$$

In the equation, $\Delta P$ is the pressure difference between the two ends of the vessel and represents the driving force behind blood flow. Viscosity ($\eta$) and blood vessel length ($l$) are not commonly altered in a healthy adult. We can also see from the equation that blood flow is directly proportional to the fourth power of vessel radius ($r^4$), which means that small variations in vessel radius translate into large changes in blood flow. In the human body, changing blood vessel radius provides an extremely effective and sensitive method of blood flow control. Peripheral resistance is the most important factor in blood flow control, because circulation to individual organs can be independently regulated even though systemic pressure may be changing.

## Getting Started

Begin by making sure you have the computer equipment and software listed in the Materials section on pp. 25–26.

1.  Quit all applications currently running on your computer.

2.  Insert the PhysioEx CD-ROM into the CD-ROM drive and keep PhysioEx in the drive during the entire time you use the program.

3.  Follow the instructions below for your computer type.

**Windows 95/98**
The program will launch automatically by loading the CD-ROM into the drive.
If autorun is disabled on your machine follow these instructions:
1.  Double-click the My Computer icon on the Windows desktop.
2.  Double-click the PhysioEx CD icon.
3.  Look for the icon that looks like a tiny movie projector and double-click it.

**Macintosh**
Double-click the PhysioEx CD icon that appears in the window on your desktop to launch the program.

## Vessel Resistance

Imagine for a moment that you are one of the first cardiovascular researchers interested in the physics of blood flow. Your first task as the principal investigator for this project is to plan an effective experimental design simulating a simple fluid pumping system that can be related to the mechanics of the cardiovascular system. The initial phenomenon you study is how fluids, including blood, flow through tubes or blood vessels. Questions you might ask include:

1.  What role does pressure play in the flow of fluid?

2.  How does peripheral resistance affect fluid flow?

The equipment required to solve these and other questions has already been designed for you in the form of a computerized simulation, which frees you to focus on the logic of the experiment. The first part of the computer simulation indirectly investigates the effects of pressure, vessel radius, viscosity, and vessel length on fluid flow. The second part of the experiment will explore the effects of several variables on the output of a single-chamber pump. Follow the specific guidelines in the exercise for collecting data. As you do so, also try to imagine alternate methods of achieving the same experimental goal.

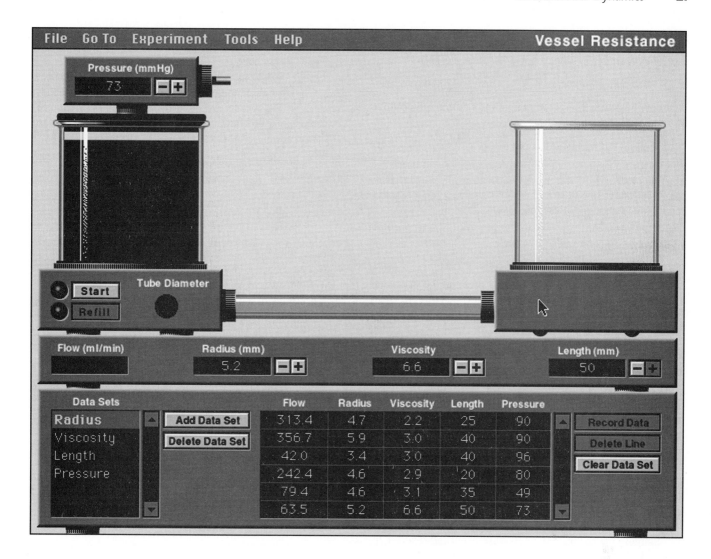

**Figure 33.1** **Opening screen of the Vessel Resistance experiment.**

Choose **Cardiovascular Dynamics** from the main menu. The opening screen will appear in a few seconds (Figure 33.1).

The primary features on the screen when the program starts are a pair of glass beakers perched atop a simulated electronic device called the *equipment control unit,* which is used to set experiment parameters and to operate the equipment. When the **Start** button (beneath the left beaker) is clicked, the simulated blood flows from the left beaker (source) to the right beaker (destination) through the connecting tube. Clicking the **Refill** button refills the source beaker after an experimental trial. Experimental parameters can be adjusted by clicking the plus (+) or minus (−) buttons to the right of each display window.

The equipment in the lower part of the screen is called the *data collection unit.* This equipment records and displays data you accumulate during the experiments. The data set for the first experiment (Radius) is highlighted in the **Data Sets** window. You can add or delete a data set by clicking the appropriate button to the right of the **Data Sets** window. The **Record Data** button at the lower right part of the screen activates automatically after an experimental trial. Clicking the **Delete Line** or **Clear Data Set** buttons erases any data you want to delete.

You will record the data you accumulate in the experimental values grid in the lower middle part of the screen.

**A c t i v i t y :**
## Studying the Effect of Flow Tube Radius on Fluid Flow

Our initial study will examine the effect of flow tube radius on fluid flow.[1]

1.   Conduct the initial equipment setup.

The **Radius** line in the data collection unit should be highlighted in bright blue. If it is not, choose it by clicking the **Radius** line. The data collection unit will now record flow variations due to changing flow tube radius.

If the data grid is not empty, click **Clear Data Set** to discard all previous values.

If the left beaker is not full, click **Refill.**

2.   Adjust the flow tube radius to 1.5 mm and the viscosity to 1.0 by clicking the appropriate plus or minus button. During the course of this part of the experiment, maintain the other experiment conditions at:

100 mm Hg driving pressure (top left)

50 mm flow tube length (middle right)

3.   Click **Start** and watch the fluid move into the right beaker. (Fluid moves slowly under some conditions—be patient!) Pressure (currently set to 100 mm Hg) propels fluid from the left beaker to the right beaker through the flow tube. The flow rate is displayed in the Flow window when the left beaker has finished draining. Now click **Record Data** to display the flow rate and experiment parameters in the experimental values grid (and retain the data in the computer's memory for printing and saving). Click **Refill** to replenish the left beaker.

4.   Increase the radius in 0.5 mm increments and repeat step 3 until the maximum radius (6.0 mm) is achieved. Be sure to click **Record Data** after each fluid transfer. If you make an error and want to delete a data value, click the data line in the experimental values grid and then click **Delete Line.**

[1] If you need help identifying any piece of equipment, choose Balloons On from the Help menu and move the mouse pointer onto any piece of equipment visible on the computer's screen. As the pointer touches the object, a pop-up window appears identifying the equipment. To close the pop-up window, move the mouse pointer away from the equipment. Repeat the process for all equipment on the screen until you feel confident with it. When finished, choose Balloons On again to turn off this help feature.

5.   View your data graphically by choosing **Plot Data** from the **Tools** menu. Choose **Radius** as the data set to be graphed, and then use the slider bars to select the radius data to be plotted on the X-axis and the flow data to be plotted on the Y-axis. Click **Draw It** to generate the graph. You can highlight individual data points by clicking a line in the data grid. When you are finished, click the close box at the top of the plot window.

What happened to fluid flow as the radius of the flow tube was increased?

_____

_____

_____

Because fluid flow is proportional to the fourth power of the

radius, _____ changes in tube radius cause

_____ changes in fluid flow.

Is the relationship between fluid flow and flow tube radius

linear or exponential? _____

In this experiment, a simulated motor changes the diameter of the flow tube. Explain how our blood vessels alter blood flow.

_____

_____

_____

_____

After a heavy meal when we are relatively inactive, we might expect blood vessels in the skeletal muscles to be somewhat

_____, whereas blood vessels in the digestive

organs are probably _____. ■

## Activity:
# Studying the Effect of Viscosity on Fluid Flow

With a viscosity of 3 to 4, blood is much more viscous than water (1.0 viscosity). Although viscosity is altered by factors such as dehydration and altered blood cell numbers, a body in homeostatic balance has a relatively stable blood consistency. Nonetheless it is useful to examine the effects of viscosity changes on fluid flow, because we can then predict what might transpire in the human cardiovascular system under certain homeostatic imbalances.

1. Set the starting conditions as follows:

- 100 mm Hg driving pressure
- 5.0 mm flow tube radius
- 1.0 viscosity
- 50 mm flow tube length

2. Click the **Viscosity** data set in the data collection unit. (This action prepares the experimental values grid to record the viscosity data.)

3. **Refill** the left beaker if you have not already done so.

4. Click **Start** to begin the experiment. After all the fluid has drained into the right beaker, click **Record Data** to record this data point, and then click **Refill** to replenish the left beaker.

5. In 1.0 unit increments, increase the fluid viscosity and repeat step 4 until the maximum viscosity (10.0) is reached.

6. View your data graphically by choosing **Plot Data** from the **Tools** menu. Choose **Viscosity** as the data set to be graphed, and then use the slider bars to select the viscosity data to be plotted on the X-axis and the flow data to be plotted on the Y-axis. Click **Draw It** to generate the graph. You can highlight individual data points by clicking a line in the data grid. When finished, click the close box at the top of the plot window.

How does fluid flow change as viscosity is modified?

_____

_____

_____

Is fluid flow versus viscosity an inverse or direct relationship?

_____

How does the effect of viscosity compare with the effect of radius on fluid flow?

_____

_____

Predict the effect of anemia (e.g., fewer red blood cells than normal) on blood flow.

_____

_____

What might happen to blood flow if we increased the number of blood cells?

_____

_____

_____

Explain why changing blood viscosity would or would not be a reasonable method for the body to control blood flow.

_____

_____

_____

_____

Blood viscosity would _____ in conditions of

dehydration, resulting in _____ blood flow. ■

## Activity:
# Studying the Effect of Flow Tube Length on Fluid Flow

With the exception of the normal growth that occurs until the body reaches full maturity, blood vessel length does not significantly change. By conducting this part of the exercise, we will be able to understand the physical relationship between vessel length and blood movement by observing how blood flow changes in flow tubes (vessels) of constant radius, but of different lengths.

1. Set the starting conditions as follows:

- 100 mm Hg driving pressure
- 5.0 mm flow tube radius
- 3.5 viscosity
- 10 mm flow tube length

2. Click the **Length** data set in the data collection unit. (This action prepares the experimental values grid to display the length data.)

3.  **Refill** the left beaker if you have not already done so.

4.  Click **Start** to begin the experiment. After all the fluid has drained into the right beaker, click **Record Data** to record this data point, and then click **Refill** to refill the left beaker.

5.  In 5 mm increments, increase the flow tube length by clicking the plus button next to the length window and repeat step 4 until the maximum length (50 mm) has been reached.

6.  View your data graphically by choosing **Plot Data** from the **Tools** menu. Choose **Length** as the data set to be graphed, and then use the slider bars to select the length data to be plotted on the X-axis and the flow data to be plotted on the Y-axis. Click **Draw It** to generate the graph. You can highlight individual data points by clicking a line in the grid. When finished, click the close box at the top of the plot window.

How does flow tube length affect fluid flow?

_____

_____

_____

Explain why altering blood vessel length would or would not be a good method of controlling blood flow in the body.

_____

_____

_____

_____ ■

### Activity:
# Studying the Effect of Pressure on Fluid Flow

The pressure difference between the two ends of a blood vessel is the driving force behind blood flow. In comparison, our experimental setup pressurizes the left beaker, thereby providing the driving force that propels fluid through the flow tube to the right beaker. You will examine the effect of pressure on fluid flow in this part of the experiment.

1.  Set the starting conditions as follows:

- 25 mm Hg driving pressure
- 5.0 mm flow tube radius
- 3.5 viscosity
- 50 mm flow tube length

2.  Click the **Pressure** data set in the data collection unit. (This action prepares the experimental values grid for the pressure data.)

3.  **Refill** the left beaker if you have not already done so.

4.  Click **Start** to begin the experiment. After all the fluid has moved into the right beaker, click **Record Data** to record this data point. Click **Refill** to refill the left beaker.

5.  In 25 mm Hg increments, increase the driving pressure by clicking the plus button next to the pressure window and repeat step 4 until the maximum pressure (225 mm Hg) has been reached.

6.  View your data graphically by choosing **Plot Data** from the **Tools** menu. Choose **Pressure** as the data set to be graphed, and then use the slider bars to select the pressure data to be plotted on the X-axis and the flow data to be plotted on the Y-axis. Click **Draw It** to generate the graph. You can highlight individual data points by clicking a line in the data grid. When finished, click the close box at the top of the plot window.

How does driving pressure affect fluid flow?

_____

_____

_____

How does this plot differ from the others?

_____

_____

_____

Although changing pressure could be used as a means of blood flow control, explain why this approach would not be as effective as altering blood vessel radius.

_____

_____

_____

_____ ■

## Pump Mechanics

In the human body, the heart beats approximately 70 strokes each minute. Each heart beat consists of a filling interval during which blood moves into the chambers of the heart, and an ejection period when blood is actively pumped into the great arteries. The pumping activity of the heart can be described in terms of the phases of the cardiac cycle. Heart chambers fill during **diastole** (relaxation of the heart) and pump blood out during **systole** (contraction of the heart). As you can imagine, the length of time the heart is relaxed is one factor that determines the amount of blood within the heart at the end of the filling interval. Up to a point, increasing ventricular filling time results in a corresponding increase in ventricular volume. The volume in the ventricles at the end of diastole, just before cardiac contraction, is called the **end diastolic volume,** or **EDV.**

Blood moves from the heart into the arterial system when systolic pressure increases above the residual pressure (from the previous systole) in the great arteries leaving the heart. Although ventricular contraction causes blood ejection, the heart does not empty completely. As a result, a small quantity of blood (**end systolic volume,** or **ESV**) remains in the ventricles at the end of systole.

Because the oxygen requirements of body tissue change depending on activity levels, we would expect the **cardiac output** (amount of blood pumped by each ventricle per minute) to vary correspondingly. We can calculate the stroke volume (amount of blood pumped per contraction of each ventricle) by subtracting the end systolic volume from the end diastolic volume ($SV = EDV - ESV$). It is possible to then compute cardiac output by multiplying the stroke volume by heart rate.

The human heart is a complex four-chambered organ, consisting of two individual pumps (the right and left sides) connected together in series. The right heart pumps blood through the lungs into the left heart, which in turn delivers blood to the systems of the body. Blood then returns to the right heart to complete the circuit.

Using the Pump Mechanics part of the program you will explore the operation of a simple one-chambered pump, and apply the physical concepts gained in the simulation to the operation of either of the two pumps comprising the human heart.

In this experiment you can vary the starting and ending volumes of the pump (analogous to EDV and ESV, respectively), driving and resistance pressures, and the diameters of the flow tubes leading to and from the pump chamber. As you proceed through the exercise, try to apply the ideas of ESV, EDV, cardiac output, stroke volume, and blood flow to the on-screen simulated pump system. For example, imagine that the flow tube leading to the pump from the left represents the pulmonary veins, while the flow tube exiting the pump to the right represents the aorta. The pump would then represent the left side of the heart.

Select **Pump Mechanics** from the **Experiment** menu. The equipment for the Pump Mechanics part of the experiment will become visible within a moment (Figure 33.2).

As you discovered in the previous experiment, there are two simulated electronic control units on the computer's screen. The upper apparatus is the *equipment control unit,* which is used to adjust experiment parameters and to operate the equipment. The lower apparatus is the *data collection* unit, in which you will record the data you accumulate during the course of the experiment.

This equipment is slightly different than that used in the Vessel Resistance experiment. There are two beakers: the *source* beaker on the left and the *destination* beaker on the right. The pressure in each beaker is individually controlled by the small pressure units on top of the beakers. Between the two beakers is a simple pump, which can be thought of as one

side of the heart, or even as a single ventricle (e.g., left ventricle). The left beaker and flow tube are analogous to the venous side of human blood flow, while arterial circulation is simulated by the right flow tube and beaker. One-way valves in the flow tubes supplying the pump ensure fluid movement in one direction—from the left beaker into the pump, and then into the right beaker. If you imagine that the pump represents the left ventricle, then think of the valve to the left of the pump as the bicuspid valve and the valve to the right of the pump as the aortic semilunar valve. The pump is driven by a pressure unit mounted on its cap. An important distinction between the pump's pressure unit and the pressure units atop the beakers is that the pump only delivers pressure during its downward stroke. Upward pump strokes are driven by pressure from the left beaker (the pump does not exert any resistance to flow from the left beaker during pump filling). In contrast, pressure in the right beaker works against the pump pressure, which means that the net pressure driving the fluid into the right beaker is calculated (automatically) by subtracting the right beaker pressure from the pump pressure. The resulting pressure difference between the pump and the right beaker is displayed in the experimental values grid in the data collection unit as **Pres.Dif.R.**

Clicking the **Auto Pump** button will cycle the pump through the number of strokes indicated in the Max. strokes window. Clicking the **Single** button cycles the pump through one stroke. During the experiment, the pump and flow rates are automatically displayed when the number of pump strokes is 5 or greater. The radius of each flow tube is individually controlled by clicking the appropriate button. Click the plus button to increase flow tube radius or the minus button to decrease flow tube radius.

The pump's stroke volume (the amount of fluid ejected in one stroke) is automatically computed by subtracting its ending volume from the starting volume. You can adjust starting and ending volumes, and thereby stroke volume, by clicking the appropriate plus or minus button in the equipment control unit.

The data collection unit records and displays data you accumulate during the experiments. The data set for the first experiment (**Rad.R.,** which represents right flow tube radius) is highlighted in the **Data Sets** window. You can add or delete a data set by clicking the appropriate button to the right of the **Data Sets** window. Clicking **Delete Data Set** will erase the data set itself, including all the data it contained. The **Record Data** button at the right edge of the screen activates automatically after an experimental trial. When clicked, the **Record Data** button displays the flow rate data in the experimental values grid and saves it in the computer's memory. Clicking the **Delete Line** or **Clear Data Set** button erases any data you want to delete.

You will record the data you accumulate in the experimental values grid in the lower middle part of the screen.

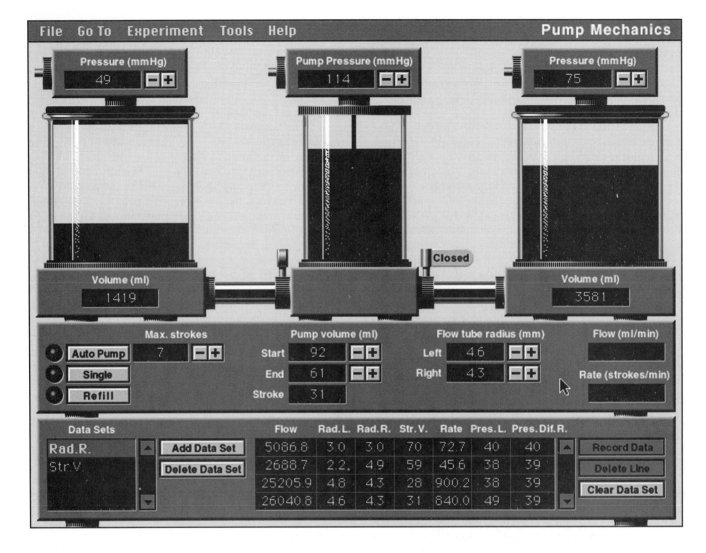

**Figure 33.2 Opening screen of the Pump Mechanics experiment.**

Activity:
## Studying the Effect of Radius on Pump Activity

Although you will only be manipulating the radius of the right flow tube in this part of the exercise, try to predict the consequence of altering the left flow tube radius as you collect the experimental data. (Remember that the left flow tube simulates the pulmonary veins and the right flow tube simulates the aorta.)

1.   Conduct the initial equipment setup.

Click the **Rad.R.** data set to activate it. The data collection unit is now ready to record flow variations due to changing flow tube radius.

If the experimental values grid is not empty, click **Clear Data Set** to discard all previous values.

If the left beaker is not full, click **Refill.**

2.   Adjust the **right** flow tube radius to 2.5 mm, and the **left** flow tube radius to 3.0 mm by clicking and holding the ap-

propriate button. During the entire part of the radius experiment, maintain the other experiment conditions at:

- 40 mm Hg left beaker pressure (this pressure drives fluid into the pump, which offers no resistance to filling)
- 120 mm Hg pump pressure (pump pressure is the driving force that propels fluid into the right beaker)
- 80 mm Hg right beaker pressure (this pressure is the resistance to the pump's pressure)
- 120 ml starting volume (analogous to EDV)
- 50 ml ending volume (analogous to ESV)
- 10 strokes in the Max. strokes window

Notice that the displayed 70 ml stroke volume is automatically calculated. Before starting, click the **Single** button one or two times and watch the pump action.

To be sure you understand how this simple mechanical pump can be thought of as a simulation of the human heart, complete the following statements by circling the correct term within the parentheses:

a.   When the piston is at the bottom of its travel, the volume remaining in the pump is analogous to (EDV, ESV) of the heart.

b.   The amount of fluid ejected into the right beaker by a single pump cycle is analogous to (stroke volume, cardiac output) of the heart.

c.   The volume of blood in the heart just before systole is called (EDV, ESV), and is analogous to the volume of fluid present in the simulated pump when it is at the (top, bottom) of its stroke.

3.   Click the **Auto Pump** button in the equipment control unit to start the pump. After the 10 stroke volumes have been delivered, the Flow and Rate windows will automatically display the experiment results. Now click **Record Data** to display the figures you just collected in the experimental values grid. Click **Refill** to replenish the left beaker.

4.   Increase the *right* flow tube radius in 0.5 mm increments and repeat step 3 above until the maximum radius (6.0 mm) is achieved. Be sure to click **Record Data** after each trial.

5.   When you have completed the experiment, view your data graphically by choosing **Plot Data** from the **Tools** menu. Choose **Rad.R.** as the data set to be graphed, and then use the slider bars to select the Rad.R. data to be plotted on the X-axis and the flow data to be plotted on the Y-axis. Click **Draw It** to generate the graph. You can highlight individual data points by clicking a line in the data grid.

The total flow rates you just determined depend on the flow rate into the pump from the left and on the flow rate out of the pump toward the right. Consequently, the shape of the plot is different than what you might predict after viewing the vessel resistance radius graph.

Try to explain why this graph differs from the radius plot in the Vessel Resistance experiment. Remember that the flow rate into the pump did not change, whereas the flow rate out of the pump varied according to your radius manipulations. When you have finished, click the close box at the top of the plot window.

_____

_____

_____

_____

Complete the following statements by circling the correct term within the parentheses.

a.   As the right flow tube radius is increased, fluid flow rate (increases, decreases). This is analogous to (dilation, constriction) of blood vessels in the human body.

b.   Even though the pump pressure remains constant, the pump rate (increases, decreases) as the radius of the right flow tube is increased. This happens because the resistance to fluid flow is (increased, decreased).

Apply your observations of the simulated mechanical pump to complete the following statements about human heart function. If you are not sure how to formulate your response, use the simulation to arrive at an answer. Circle the correct term within the parentheses.

c.   The heart must contract (more, less) forcefully to maintain cardiac output if the resistance to blood flow in the vessels exiting the heart is increased.

d.   Increasing the resistance (e.g., a constriction) of the blood vessels entering the heart would (increase, decrease) the time needed to fill the heart chambers.

What do you think would happen to the flow rate and the pump rate if the left flow tube radius is changed (either increased or decreased)?

_____

_____

_____  ∎

## Activity:
## Studying the Effect of Stroke Volume on Pump Activity

Whereas the heart of a person at rest pumps about 60% of the blood in its chambers, 40% of the total amount of blood remains in the chambers after systole. The 60% of blood ejected by the heart is called the stroke volume, and is the difference between EDV and ESV. Even though our simple pump in this experiment does not work exactly like the human heart, you can apply the concepts to basic cardiac function. In this experiment, you will examine how the activity of the simple pump is affected by changing the pump's starting and ending volumes.

1.   Click the **Str.V.** (stroke volume) data set to activate it. If the experimental values grid is not empty, click **Clear Data Set** to discard all previous values. If the left beaker is not full, click **Refill.**

2.   Adjust the stroke volume to 10 ml by setting the Start volume (EDV) to 120 ml and the End volume (ESV) to 110 ml (stroke volume = start volume − end volume). During the entire stroke volume part of the experiment, keep the other experimental conditions at:

*   40 mm Hg left beaker pressure
*   120 mm Hg pump pressure
*   80 mm Hg right beaker pressure
*   3.0 mm left and right flow tube radius
*   10 strokes in the Max. strokes window

3.   Click the **Auto Pump** button to start the experiment. After 10 stroke volumes have been delivered, the Flow and Rate windows will display the experiment results given the current parameters. Click **Record Data** to display the figures you just collected in the experimental values grid. Click **Refill** to replenish the left beaker.

4.   Increase the stroke volume in 10 ml increments (by decreasing the End volume) and repeat step 3 until the maximum stroke volume (120 ml) is achieved. Be sure to click the **Record Data** button after each trial. Watch the pump action during each stroke to see how you can apply the concepts of starting and ending pump volumes to EDV and ESV of the heart.

5.   View your data graphically by choosing **Plot Data** from the **Tools** menu. Choose **Str.V.** as the data set to be graphed, and then use the slider bars to select the Str.V. data to be plotted on the X-axis and the pump rate data to be plotted on the Y-axis. Click **Draw It** to generate the graph and then answer the following questions. When you have finished, click the close box at the top of the plot window.

What happened to the pump's rate as its stroke volume was increased?

_____

_____

_____

Using your simulation results as a basis for your answer, explain why an athlete's resting heart rate might be lower than that of the average person.

_____

_____

_____

Applying the simulation outcomes to the human heart, predict the effect of increasing the stroke volume on cardiac output (at any given rate).

_____

_____

_____

When heart rate is increased, the time of ventricular filling is (circle one: increased, decreased), which in turn (increases, decreases) the stroke volume.

What do you think might happen to the pressure in the pump during filling if the valve in the right flow tube became leaky? (Remember that the pump offers no resistance to filling.)

_____

_____

Applying this concept to the human heart, what might occur in the left heart and pulmonary blood vessels if the aortic valve became leaky?

_____

_____

What might occur if the aortic valve became slightly constricted?

_____

_____

In your simulation, increasing the pressure in the right beaker is analogous to the aortic valve becoming (circle one: leaky, constricted). ∎

### A c t i v i t y :
## Studying Combined Effects

In this section, you will set up your own experimental conditions to answer the following questions. Carefully examine each question and decide how to set experiment parameters to arrive at an answer. You should try to set up your own experimental conditions to formulate a response, but you can examine your previously collected data at any time if you need additional information. As a rule of thumb, record several data points for each question as evidence for your answer (unless the question calls for a single pump stroke).

   Click the **Add Data Set** button in the data collection unit. Next, create a new data set called Combined. Your newly created data set will be displayed beneath Str.V. Now click the **Combined** line to activate the data set. As you collect the supporting data for the following questions, be sure to click **Record Data** each time you have a data point you wish to keep for your records.

How is the flow rate affected when the right flow tube radius is kept constant (at 3.0 mm) and the left flow tube radius is modified (either up or down)?

_____

_____

_____

_____

How does decreasing left flow tube radius affect pump chamber filling time? Does it affect pump chamber emptying?

_____

_____

_____

You have already examined the effect of changing the pump's end volume as a way of manipulating stroke volume. What happens to flow and pump rate when you keep the end volume constant and alter the start volume to manipulate stroke volume?

_____

_____

_____

_____

Try manipulating the pressure delivered to the left beaker. How does changing the left beaker pressure affect flow rate? (This change would be similar to changing pulmonary vein pressure.)

_____

_____

_____

If the left beaker pressure is decreased to 10 mm Hg, how is pump-filling time affected?

_____

_____

_____

What happens to the pump rate if the filling time is shortened?

_____

_____

What happens to fluid flow when the right beaker pressure equals the pump pressure?

_____

_____

_____ ■

## Activity:
# Studying Compensation

Click the **Add Data Set** button in the data collection unit. Next, create a new data set called **Comp.** Your newly created data set will be displayed beneath Combined. Now click the **Comp. Line** to activate the data set. As you collect the supporting data for the following questions, be sure to click **Record Data** each time you have a data point you wish to keep for your records.

Adjust the experimental conditions to the following:

- 40 mm Hg left beaker pressure
- 120 mm Hg pump pressure
- 80 mm Hg right beaker pressure
- 3.0 mm left and right flow tube radius
- 10 strokes in the Max. strokes window
- 120 ml Starting volume
- 50 ml Ending volume

Click **Auto Pump** and then record your flow rate data. Let's declare the value you just obtained to be the "normal" flow rate for the purpose of this exercise.

Now decrease the right flow tube radius to 2.5 mm, and run another trial. How does this flow rate compare with "normal"?

_____

_____

Leave the right flow tube radius at 2.5 mm radius and try to adjust one or more other conditions to return flow to "normal." Think logically about what condition(s) might compensate for a decrease in flow tube radius. How were you able to accomplish this? (Hint: There are several ways.)

_____

_____

_____

Decreasing the right flow tube radius is similar to a partial (circle one: leakage, blockage) of the aortic valve or (increased, decreased) resistance in the arterial system.

Explain how the human heart might compensate for such a condition.

_____

_____

_____

_____

If we wanted to increase (or decrease) blood flow to a particular body system (e.g., digestive), would it be better to adjust heart rate or blood vessel diameter? Explain.

_____

_____

_____

_____

Complete the following statements by circling the correct response. (If necessary, use the pump simulation to help you with your answers.)

a.   If we decreased overall peripheral resistance in the human body (as in an athlete), the heart would need to generate (more, less) pressure to deliver an adequate amount of blood flow, and arterial pressure would be (higher, lower).

b.   If the diameter of the arteries of the body were partly filled with fatty deposits, the heart would need to generate (more, less) force to maintain blood flow, and pressure in the arterial system would be (higher, lower) than normal. ■

Activity:
# Using Your Data

## Save Your Data

When you have finished all parts of the exercise, you can save your data set for later review and additional analysis.

1.   Select **Save Data to File** from the **File** menu.

2.   When the dialog box opens, click a drive and/or directory location.

3.   Type a file name into the **File name** box (your instructor may provide you with a name).

4.   Click **Save** to complete the save. If you want to review previously saved data, select **Open Data File** from the File menu and choose your file.

## Printing Your Data

If your instructor requires a printed copy of your experiment results, choose **Print** from the **File** menu.

1.   Type your name into the space provided.

2.   When you are ready to print, click OK. ■

# Frog Cardiovascular Physiology: Computer Simulation

## Objectives

1. To list the properties of cardiac muscle as automaticity and rhythmicity and define each.

2. To explain the statement "Cardiac muscle has an intrinsic ability to beat."

3. To compare the relative length of the refractory period of cardiac muscle with that of skeletal muscle, and explain why it is not possible to tetanize cardiac muscle.

4. To define *extrasystole* and to explain at what point in the cardiac cycle (and on an ECG tracing) an extrasystole can be induced.

5. To describe the effect of the following on heart rate: cold, heat, vagal stimulation, pilocarpine, digitalis, atropine, epinephrine, and potassium, sodium, and calcium ions.

6. To define *vagal escape* and discuss its value.

7. To define *ectopic pacemaker*.

## Materials

*Minimum equipment required:*

**Windows:**

❑ 486/66MHz or better strongly recommended

❑ Windows 95/98 or higher recommended

❑ SVGA display (256 colors at 640 x 480)

❑ Minimum 16 MB available RAM

❑ Double-speed CD-ROM drive (quad-speed or higher recommended)

❑ Sound card

❑ Speakers or headphones

❑ Printer

**Macintosh:**

❑ 68040 or Power Macintosh strongly recommended

❑ System 7.1 or higher

❑ Minimum 16 MB available RAM

❑ 13-inch or larger color monitor (640 x 480 resolution)

❑ Double-speed CD-ROM drive (quad-speed or higher recommended)

❑ Printer

**Software:**

❑ Benjamin/Cummings PhysioEx CD-ROM—Frog Cardiovascular Physiology module

nvestigation of human cardiovascular physiology is very interesting, but many areas obviously do not lend themselves to experimentation. It would be tantamount to murder to inject a human subject with various drugs to observe their effects on heart activity, or to expose the human heart in order to study the length of its refractory period. However, this type of investigation can be done on frogs or computer simulations, and provides valuable data, because the physiologic mechanisms in these animals, or programmed into the computer simulation, are similar if not identical to those in humans.

In this exercise, you will conduct the cardiac investigations just mentioned.

# Special Electrical Properties of Cardiac Muscle: Automaticity and Rhythmicity

Cardiac muscle differs from skeletal muscle both functionally and in its fine structure. Skeletal muscle must be electrically stimulated to contract. In contrast, heart muscle can and does depolarize spontaneously in the absence of external stimulation. This property, called **automaticity,** is due to plasma membranes that have reduced permeability to potassium ions, but still allow sodium ions to slowly leak into the cells. This leakage causes the muscle cells to slowly depolarize until the action potential threshold is reached and *fast calcium channels* open, allowing $Ca^{2+}$ entry from the extracellular fluid. Shortly thereafter, contraction occurs.

Additionally, the spontaneous depolarization-repolarization events occur in a regular and continuous manner in cardiac muscle, a property referred to as **rhythmicity.**

In the following experiment, you will observe these properties of cardiac muscle in a computer simulation. Additionally, your instructor may demonstrate this procedure using a real frog.

# Getting Started

Begin by making sure you have the computer equipment and software listed in the Materials section on page 37.

1.   Quit all applications currently running on your computer.

2.   Insert the PhysioEx CD-ROM into the CD-ROM drive and keep PhysioEx in the drive during the entire time you use the program.

3.   Follow the instructions below for your computer type.

### Windows 95/Windows 98
The program will launch automatically after you load the CD-ROM into the drive.
If autorun is disabled on your machine follow these instructions:
1.   Double-click the My Computer icon on the Windows desktop.
2.   Double-click the PhysioEx CD icon.
3.   Look for the icon that looks like a tiny movie projector and double-click it.

### Macintosh
Double-click the PhysioEx CD icon that appears in the window on your desktop to launch the program.

# Baseline Frog Heart Activity

The heart's effectiveness as a pump is dependent both on intrinsic (within the heart) and extrinsic (external to the heart) controls. In this experiment, you will investigate some of these factors.

The nodal system, in which the "pacemaker" imposes its depolarization rate on the rest of the heart, is one intrinsic factor that influences the heart's pumping action. If its impulses fail to reach the ventricles (as in heart block), the ventricles continue to beat but at their own inherent rate, which is much slower than that usually imposed on them. Although heart contraction does not depend on nerve impulses, its rate can be modified by extrinsic impulses reaching it through the autonomic nerves. Additionally, cardiac activity is modified by various chemicals, hormones, ions, and metabolites. The effects of several of these chemical factors are examined in the next experimental series.

The frog heart has two atria and a single, incompletely divided ventricle. The pacemaker is located in the sinus venosus, an enlarged region between the venae cavae and the right atrium. The SA node of mammals may have evolved from the sinus venosus.

Choose Frog Cardiovascular Physiology from the main menu. The opening screen will appear in a few seconds (Figure 35B.1). When the program starts, you will see a tracing of the frog's heart beat on the *oscilloscope display* in the upper right part of the screen. Because the simulation automatically adjusts itself to your computer's speed, you may not see the heart tracing appear in real-time. If you want to increase the speed of the tracing (at the expense of tracing quality), click the **Tools** menu, choose **Modify Display,** and then select **Increase Speed.**

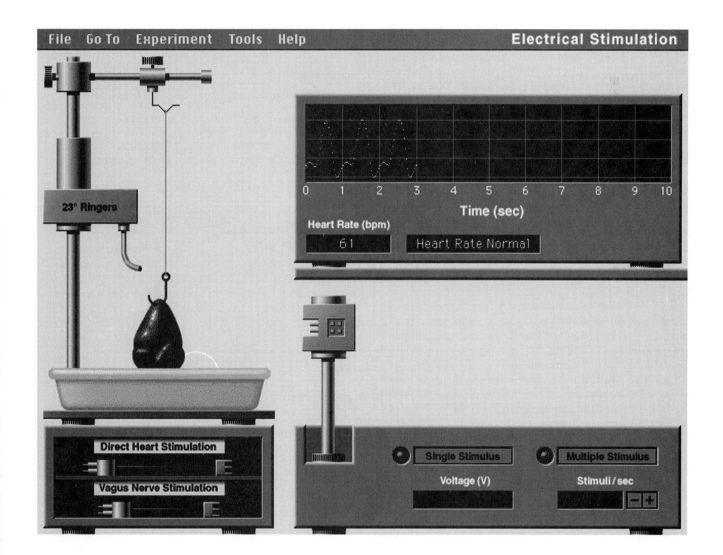

**Figure 35B.1   Opening screen of the Electrical Stimulation experiment.**

The oscilloscope display shows the ventricular contraction rate in the Heart rate window. The *heart activity window* to the right of the Heart Rate display provides the following messages:

• Heart Rate Normal—displayed when the heart is beating under resting conditions.

• Heart Rate Changing—displayed when the heart rate is increasing or decreasing.

• Heart Rate Stable—displayed when the heart rate is steady, but higher or lower than normal. For example, if you applied a chemical that increased heart rate to a stable but higher-than-normal rate, you would see this message.

The *electrical stimulator* is below the oscilloscope display. In the experiment, clicking **Single Stimulus** delivers a single electrical shock to the frog heart. Clicking **Multiple Stimulus** delivers repeated electrical shocks at the rate indicated in the Stimuli/sec window just below the Multiple Stimulus button. When the **Multiple Stimulus** button is clicked, it changes to a **Stop Stimulus** button that allows you to stop electrical stimulation as desired. Clicking the (+) or (−) buttons next to the Stimuli/sec window adjusts the stimulus rate. The voltage delivered when Single Stimulus or Multiple Stimulus is clicked is displayed in the Voltage window. The simulation automatically adjusts the voltage for the experiment. The post-like apparatus extending upward from the electrical stimulator is the *electrode holder* into which you will drag-and-drop electrodes from the supplies cabinet in the bottom left corner of the screen.

The left side of the screen contains the apparatus that sustains the frog heart. The heart has been lifted away from the body of the frog by a hook passed through the apex of the heart. Although the frog can not be seen because it is in the dissection tray, its heart has not been removed from its circulatory system. A thin string connects the hook in the heart to the force transducer at the top of the support bracket. As the heart contracts, the string exerts tension on the force transducer that converts the contraction into the oscilloscope tracing. The slender white strand extending from the heart toward the right side of the dissection tray is the vagus nerve. In the simulation, room temperature frog Ringer's solution continuously drips onto the heart to keep it moist and responsive so that a regular heart beat is maintained.

The two electrodes you will use during the experiment are located in the supplies cabinet beneath the dissection tray. The Direct Heart Stimulation electrode is used to stimulate the ventricular muscle directly. The Vagus Nerve Stimulation electrode is used to stimulate the vagus nerve. To position either electrode, click and drag the electrode to the two-pronged plug in the electrode holder and then release the mouse button.

### Activity:
## Recording Baseline Frog Heart Activity

1. Before beginning to stimulate the frog heart experimentally, watch several heart beats. Be sure you can distinguish atrial and ventricular contraction (Figure 35B.2a).

2. Record the number of ventricular contractions per minute displayed in the Heart Rate window under the oscilloscope.

_____ bpm (beats per minute) ■

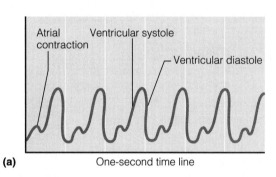

**(a)**    One-second time line

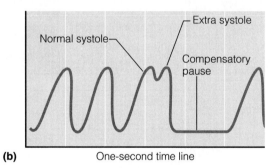

**(b)**    One-second time line

**Figure 35B.2  Recording of contractile activity of a frog heart.** (a) Normal heartbeat. (b) Induction of an extrasystole.

Activity:

# Investigating the Refractory Period of Cardiac Muscle

In Exercise 16B you saw that repeated rapid stimuli could cause skeletal muscle to remain in a contracted state. In other words, the muscle could be tetanized. This was possible because of the relatively short refractory period of skeletal muscle. In this experiment you will investigate the refractory period of cardiac muscle and its response to stimulation.

1. Click and hold the mouse button on the Direct Heart Stimulation electrode and drag it to the electrode holder.

2. Release the mouse button to lock the electrode in place. The electrode will touch the ventricular muscle tissue.

3. Deliver single shocks by clicking **Single Stimulus** at each of the following times. You may need to practice to acquire the correct technique.

- Near the beginning of ventricular contraction
- At the peak of ventricular contraction
- During the relaxation part of the cycle

Watch for *extrasystoles,* which are extra beats that show up riding on the ventricular contraction peak. Also note the compensatory pause, which allows the heart to get back on schedule after an extrasystole. (Figure 35B.2b).

During which portion of the cardiac cycle was it possible to induce an extrasystole?

_____

4. Attempt to tetanize the heart by clicking **Multiple Stimulus.** Electrical shocks will be delivered to the muscle at a rate of 20 stimuli/sec. What is the result?

_____

_____

Considering the function of the heart, why is it important that the heart muscle cannot be tetanized?

_____

_____

5. Click **Stop Stimulus** to stop the electrical stimulation. ■

Activity:

# Examining the Effect of Vagus Nerve Stimulation

The vagus nerve carries parasympathetic impulses to the heart, which modify heart activity.

1. Click the Direct Heart Stimulation electrode to return it to the supplies cabinet.

2. Click and drag the Vagus Nerve Stimulation electrode to the electrode holder.

3. Release the mouse button to lock the electrode in place. The vagus nerve will automatically be draped over the electrode contacts.

4. Adjust the stimulator to 50 stimuli/sec by clicking the (+) or (−) buttons.

5. Click **Multiple Stimulus.** Allow the vagal stimulation to continue until the heart stops momentarily and then begins to beat again (vagal escape) and then click **Stop Stimulus.**

What is the effect of vagal stimulation on heart rate?

_____

_____

_____

The phenomenon of vagal escape demonstrates that many factors are involved in heart regulation and that any deleterious factor (in this case, excessive vagal stimulation) will be overcome, if possible, by other physiological mechanisms such as activation of the sympathetic division of the autonomic nervous system (ANS). ■

# Assessing Physical and Chemical Modifiers of Heart Rate

Now that you have observed normal frog heart activity, you will have an opportunity to investigate the effects of various modifying factors on heart activity. After removing the agent in each activity, allow the heart to return to its normal rate before continuing with the testing.

Choose **Modifiers of Heart Rate** from the Experiment menu. The opening screen will appear in a few seconds (Figure 35B.3). The appearance and functionality of the *oscilloscope display* is the same as it was in the Electrical Stimulation experiment. The *solutions shelf* above the oscilloscope display contains the chemicals you'll use to modify heart rate in the experiment. You can choose the temperature of the Ringer's solution dispensed by clicking the appropriate button in the Ringer's dispenser at the left part of the screen. The

doors to the supplies cabinet are closed during this experiment because the electrical stimulator is not used.

When you click **Record Data** in the *data control unit* below the oscilloscope, your data is stored in the computer's memory and is displayed in the data grid at the bottom of the screen; data displayed include the solution used and the resulting heart rate. If you are not satisfied with a trial, you can click **Delete Line.** Click **Clear Table** if you wish to repeat the entire experiment.

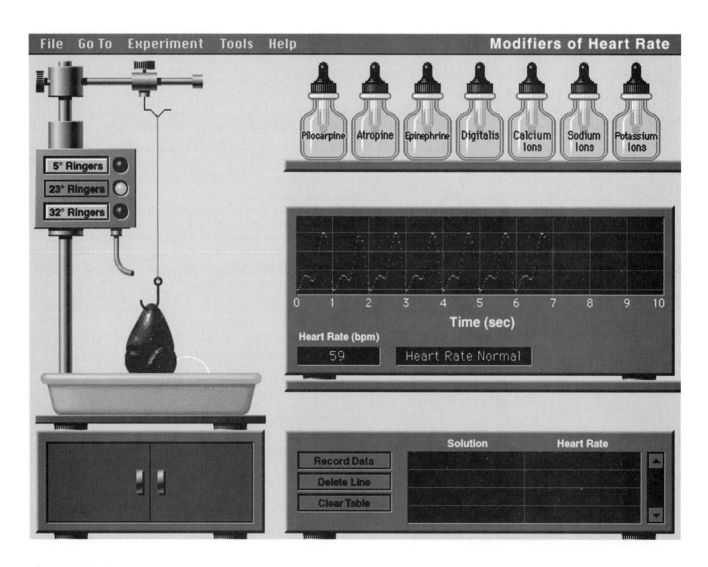

**Figure 35B.3   Opening screen of the Modifiers of Heart Rate experiment.**

Activity:
# Assessing the Effect of Temperature

1.   Click the **5°C Ringer's** button to bathe the frog heart in cold Ringer's solution. Watch the recording for a change in cardiac activity.

2.   When the heart activity window displays the message: Heart Rate Stable, click **Record Data** to retain your data in the data grid.

What change occurred with the cold (5°C) Ringer's solution?

_____

3.   Now click the **23°C Ringer's** button to flood the heart with fresh room temperature Ringer's solution.

4.   After you see the message: Heart Rate Normal in the heart activity window, click the **32°C Ringer's** button.

5.   When the heart activity window displays the message: Heart Rate Stable, click **Record Data** to retain your data.

What change occurred with the warm (32°C) Ringer's solution?

_____

Record the heart rate at the two temperatures below.

_____ bpm at 5°C; _____ bpm at 32°C

What can you say about temperature and heart rate?

_____

_____

6.   Click the **23°C Ringer's** button to flush the heart with fresh Ringer's solution. Watch the heart activity window for the message: Heart Rate Normal before beginning the next test.  ∎

Activity:
# Assessing the Effect of Pilocarpine

1.   Click and hold the mouse on the Pilocarpine dropper cap.

2.   Drag the dropper cap to a point about an inch above the heart and release the mouse.

3.   Pilocarpine solution will be dispensed onto the heart and the dropper cap will automatically return to the Pilocarpine bottle.

4.   Watch the heart activity window for the message: Heart Rate Stable, indicating that the heart rate has stabilized under the effects of pilocarpine.

5.   After the heart rate stabilizes, record the heart rate in the space provided below, and click **Record Data** to retain your data in the grid.

_____ bpm

What happened when the heart was bathed in the pilocarpine solution?

_____

6.   Click the **23°C Ringer's** button to flush the heart with fresh Ringer's solution. Watch the heart activity window for the message: Heart Rate Normal, an indication that the heart is ready for the next test.  ∎

Pilocarpine simulates the effect of parasympathetic nerve (hence, vagal) stimulation by enhancing acetylcholine release; such drugs are called parasympathomimetic drugs.

Activity:

## Assessing the Effect of Atropine

1.   Drag-and-drop the Atropine dropper cap to a point about an inch above the heart.

2.   Atropine solution will automatically drip onto the heart and the dropper cap will return to its position in the Atropine bottle.

3.   Watch the heart activity window for the message: Heart Rate Stable.

4.   After the heart rate stabilizes, record the heart rate in the space below, and click **Record Data** to retain your data in the grid.

_____ bpm

What is the effect of atropine on the heart?

_____

Atropine is a drug that blocks the effect of the neurotransmitter acetylcholine, liberated by the parasympathetic nerve endings. Do your results accurately reflect this effect of atropine?

_____

Are pilocarpine and atropine agonists or antagonists in their effects on heart activity?

_____

5.   Click the **23°C Ringer's** button to flush the heart with fresh Ringer's solution. Watch the heart activity window for the message: Heart Rate Normal, before beginning the next test. ■

Activity:

## Assessing the Effect of Epinephrine

1.   Drag-and-drop the Epinephrine dropper cap to a point about an inch above the heart.

2.   Epinephrine solution will be dispensed onto the heart and the dropper cap will return to the Epinephrine bottle.

3.   Watch the heart activity window for the message: Heart Rate Stable.

4.   After the heart rate stabilizes, record the heart rate in the space provided below, and click **Record Data** to retain your data in the grid.

_____ bpm

What happened when the heart was bathed in the epinephrine solution?

_____

Which division of the autonomic nervous system does its effect imitate?

_____

5.   Click the **23°C Ringer's** button to flush the heart with fresh Ringer's solution. Watch the heart activity window for the message: Heart Rate Normal, meaning that the heart is ready for the next test. ■

Activity:
# Assessing the Effect of Digitalis

1.  Drag-and-drop the Digitalis dropper cap to a point about an inch above the heart.

2.  Digitalis solution will automatically drip onto the heart and then the dropper will return to the Digitalis bottle.

3.  Watch the heart activity window to the right of the Heart Rate window for the message: Heart Rate Stable.

4.  After the heart rate stabilizes, record the heart rate in the space provided below, and click **Record Data** to retain your data in the grid.

_____ bpm

What is the effect of digitalis on the heart?

_____

5.  Click the **23°C Ringer's** button to flush the heart with fresh Ringer's solution. Watch the heart activity window for the message: Heart Rate Normal, then proceed to the next test. ■

Digitalis is a drug commonly prescribed for heart patients with congestive heart failure. It slows heart rate, providing more time for venous return and decreasing the workload on the weakened heart. These effects are thought to be due to inhibition of the $Na^+-K^+$ pump and enhancement of $Ca^{2+}$ entry into myocardial fibers.

Activity:
# Assessing the Effect of Various Ions

To test the effect of various ions on the heart, apply the desired solution using the following method.

1.  Drag-and-drop the Calcium Ions dropper cap to a point about an inch above the heart.

2.  Calcium Ions will automatically be dripped onto the heart and the dropper cap will return to the Calcium Ions bottle.

3.  Watch the heart activity window for the message: Heart Rate Stable.

4.  After the heart rate stabilizes, record the heart rate in the space provided below, and click **Record Data** to retain your data in the grid.

5.  Click the **23°C Ringer's** button to flush the heart with fresh Ringer's solution. Watch the heart activity window for the message: Heart Rate Normal, which means that the heart is ready for the next test.

6.  Repeat steps 1 through 5 for Sodium Ions and then Potassium Ions.

Effect of $Ca^{2+}$:

Heart rate _____ bpm

Describe the results

_____

_____

Effect of $Na^+$:

Heart rate _____ bpm

Describe the results

_____

_____

Effect of $K^+$:

Heart rate _____ bpm

Describe the results

_____

_____ ■

Potassium ion concentration is normally higher within cells than in the extracellular fluid. *Hyperkalemia* decreases the resting potential of plasma membranes, thus decreasing the force of heart contraction. In some cases, the conduction rate of the heart is so depressed that **ectopic pacemakers** (pacemakers appearing erratically and at abnormal sites in the heart muscle) appear in the ventricle, and fibrillation may occur.

Was there any evidence of premature beats in the recording of potassium ion effects?

_____

Was arrhythmia produced with any of the ions tested?

_____ If so, which?

_____ ■

**A c t i v i t y :**
# Using Your Data

## Save Your Data

When you have finished with all experiments, you can save your data set for later review and additional analysis.

1.   Select **Save Data to File** from the **File** menu.

2.   When the dialog box opens, click a drive and/or directory location.

3.   Type a file name into the File name box (your instructor may provide you with a name).

4.   Click **Save** to complete the save. If you want to review previously-saved data, select **Open Data File** from the **File** menu, and choose your file.

## Print Your Data

If your instructor requires a hard copy of your experiment results, choose **Print Data** from the **File** menu.

1.   Type your name into the provided space.

2.   When you are ready to print, click **OK.** ■

# Chemical and Physical Processes of Digestion: Computer Simulation

## Objectives

1. To list the digestive system enzymes involved in the digestion of proteins, fats, and carbohydrates; to state their site of origin; and to summarize the environmental conditions promoting their optimal functioning.

2. To recognize the variation between different types of enzyme assays.

3. To name the end products of digestion of proteins, fats, and carbohydrates.

4. To perform the appropriate chemical tests to determine if digestion of a particular food has occurred.

5. To cite the function(s) of bile in the digestive process.

6. To discuss the possible role of temperature and pH in the regulation of enzyme activity.

7. To define *enzyme, catalyst, control, substrate,* and *hydrolase.*

8. To explain why swallowing is both a voluntary and a reflex activity.

9. To discuss the role of the tongue, larynx, and gastroesophageal sphincter in swallowing.

10. To compare and contrast segmentation and peristalsis as mechanisms of propulsion.

## Materials

Part I—Enzyme Action

Minimum equipment required:

**Windows:**

- 486/66 MHz or better recommended
- Windows 95/98 recommended
- SVGA display (256 colors at 640 × 480)
- Minimum 16 MB available RAM
- Double-speed CD-ROM drive (quad-speed or higher recommended)
- Sound card
- Speakers or headphones
- Printer

**Macintosh:**

- 68040 processor or Power Macintosh recommended
- System 7.1 or higher
- Minimum 16 MB available RAM
- 13-inch or larger color monitor (640 × 480 resolution)
- Double-speed CD-ROM drive (quad-speed or higher recommended)
- Printer

**Software:**

- Benjamin/Cummings PhysioEx CD-ROM—Chemical Processes of Digestion module

Part II—Physical Processes

- Water pitcher
- Paper cups
- Stethoscope
- Alcohol swabs
- Disposable autoclave bag
- VHS tape viewing station:

    VHS tape illustrating deglutition and peristalsis (*Passage of Food through the Digestive Tract,* available from Ward's Biology)

    VCR and television or monitor set up for independent viewing of film loop by students

The digestive system is a physiological marvel, composed of finely orchestrated chemical and physical activities. The food we ingest must be broken down to its molecular form for us to get the nutrients we need, and digestion involves a complex sequence of mechanical and chemical processes designed to achieve this goal as efficiently as possible. As food passes through the gastrointestinal tract, it is progressively broken down by the mechanical action of smooth muscle and the chemical action of enzymes until most nutrients have been extracted and absorbed into the blood.

# Chemical Digestion of Foodstuffs: Enzymatic Action

Nutrients can only be absorbed when broken down into their monomer form, so food digestion is a prerequisite to food absorption. **Enzymes** are large protein molecules produced by body cells. They are biological **catalysts** that increase the rate of a chemical reaction without becoming part of the product. The digestive enzymes are hydrolytic enzymes, or **hydrolases,** which break down organic food molecules or **substrates** by adding water to the molecular bonds, thus cleaving the bonds between the subunits or monomers.

A hydrolytic enzyme is highly specific in its action. Each enzyme hydrolyzes one or, at most, a small group of substrate molecules, and specific environmental conditions are necessary for an enzyme to function optimally. For example, temperature and pH have a large effect on the degree of enzymatic hydrolysis, and each enzyme has its preferred environment.

Because digestive enzymes actually function outside the body cells in the digestive tract lumen, their hydrolytic activity can also be studied in a test tube. Such *in vitro* studies provide a convenient laboratory environment for investigating the effect of various factors on enzymatic activity.

## Getting Started

Begin by making sure you have the computer equipment and software listed in the Materials section on p. 47.

1.   Quit all applications currently running on your computer.

2.   Insert the PhysioEx CD-ROM into the CD-ROM drive and keep PhysioEx in the drive during the entire time you use the program.

3.   Follow the instructions below for your computer type.

### Windows 95/Windows 98
The program will launch automatically when you load the CD-ROM into the drive.
If autorun is disabled on your machine follow these instructions:
1.   Double-click the My Computer icon on the Windows desktop.
2.   Double-click the PhysioEx CD icon.
3.   Look for the icon that looks like a tiny movie projector and double-click it.

### Macintosh
Double-click the PhysioEx CD icon that appears in the window on your desktop to launch the program.

## Starch Digestion by Salivary Amylase

In this experiment you will investigate the hydrolysis of starch to maltose by salivary amylase, the enzyme produced by the salivary glands and secreted into the mouth. For you to be able to detect whether or not enzymatic action has occurred, you need to be able to identify the presence of these substances to determine to what extent hydrolysis has occurred. Thus, **controls** must be prepared to provide a known standard against which comparisons can be made. The controls will vary for each experiment, and will be discussed in each enzyme section in this exercise.

Starch decreases and sugar increases as digestion proceeds according to the following equation:

$$\text{Starch} + \text{water} \xrightarrow{\text{amylase}} \text{X maltose}$$

Because the chemical changes that occur as starch is digested to maltose cannot be seen by the naked eye, you need to conduct an *enzyme assay,* the chemical method of detecting the presence of digested substances. You will perform two enzyme assays on each sample. The IKI assay detects the presence of starch and the Benedict's assay tests for the presence of maltose, which is the digestion product of starch. Normally a caramel-colored solution, IKI turns blue-black in the presence of starch. Benedict's reagent is a bright blue solution that changes to green to orange to reddish-brown with increasing amounts of maltose. It is important to understand that enzyme assays only indicate the presence or absence of substances. It is up to you to analyze the results of the experiments to decide if enzymatic hydrolysis has occurred.

Choose **Chemical and Physical Processes of Digestion** from the main menu. The opening screen will appear in a few seconds (Figure 40B.1). The *solutions shelf* in the upper right part of the screen contains the substances to be used in the ex-

periment. The *incubation unit* beneath the solutions shelf contains a rack of test tube holders and apparatus needed to run the experiments. Test tubes from the *test tube washer* on the left part of the screen are loaded into the rack in the incubation unit by clicking and holding the mouse button on the first tube, and then releasing (dragging-and-dropping) it into any position in the rack. The substances in the dropper bottles on the solutions shelf are dispensed by dragging-and-dropping the dropper cap to a position over any test tube in the rack and then releasing it. During each dispensing event, five drops of solution drip into the test tube; then the dropper cap automatically returns to its position in the bottle.

Each test tube holder in the incubation unit not only supports but also allows you to boil the contents of a single test

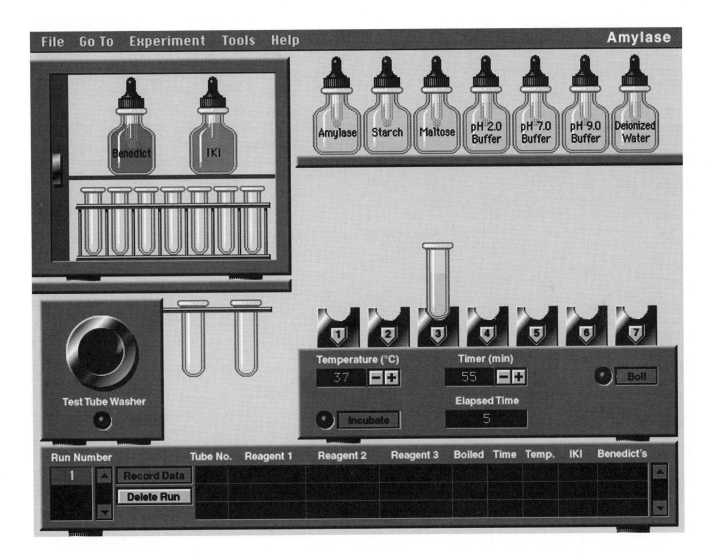

**Figure 40B.1  Opening screen of the Amylase experiment.**

tube. Clicking the numbered button at the base of a test tube holder causes that single tube to descend into the incubation unit. To boil the contents of all tubes inside the incubation unit, click **Boil.** After they have been boiled, the tubes automatically rise. You can adjust the incubation temperature for the experiment by clicking the (+) or (−) buttons next to the Temperature window. Set the incubation time by clicking the (+) or (−) buttons next to the Timer window. Clicking the **Incubate** button starts the timer and causes the entire rack of tube holders to descend into the incubation unit where the tubes will be incubated at the temperature and the time indicated. While incubating, the tubes are automatically agitated to ensure that their contents are well mixed. During the experiment elapsed time is displayed in the Elapsed Time window.

The cabinet doors in the *assay cabinet* above the test tube washer are closed at the beginning of the experiment, but they automatically open when the set time for incubation has elapsed. The assay cabinet contains the reagents and glassware needed to assay your experimental samples.

When you click the **Record Data** button in the *data control unit* at the bottom of the screen, your data is recorded in the computer's memory and displayed in the data grid at the bottom of the screen. Data displayed in the grid include the tube number, the three substances dispensed into each tube, the time and incubation temperature, and (+) or (−) marks indicating enzyme assay results and whether or not a sample was boiled. If you are not satisfied with a single run, you can click **Delete Run** to erase an experiment.

Once an experimental run is completed and you have recorded your data, discard the test tubes to prepare for a new run by dragging the used tubes to the large opening in the test tube washer. The test tubes will automatically be prepared for the next experiment.

### Activity:
## Assessing Starch Digestion by Salivary Amylase

### Amylase Incubation

1. Individually drag seven test tubes to the test tube holders in the incubation unit.

2. Prepare tubes 1 through 7 with the substances indicated in Chart 1 using the following approach.

• Click and hold the mouse button on the dropper cap of the desired substance on the solutions shelf.

• While still holding the mouse button down, drag the dropper cap to the top of the desired test tube.

• Release the mouse button to dispense the substance. The dropper cap automatically returns to its bottle.

3. When all tubes are prepared, click the number (**1**) under the first test tube. The tube will descend into the incubation unit. All other tubes should remain in the raised position.

4. Click **Boil** to boil the number 1 tube. After boiling for a few moments, the tube will automatically rise.

5. Now adjust the incubation temperature to 37°C and the timer to 60 min (compressed time) by clicking the (+) or (−) buttons.

6. Click **Incubate** to start the run. The incubation unit will gently agitate the test tube rack, evenly mixing the contents of all test tubes throughout the incubation. Notice that the computer compresses the 60-minute time period into 60 seconds of real time, so what would be a 60-minute incubation in

| Chart 1 | Salivary Amylase Digestion of Starch | | | | | | |
|---|---|---|---|---|---|---|---|
| **Tube no.** | **1** | **2** | **3** | **4** | **5** | **6** | **7** |
| **Additives** | Amylase Starch pH 7.0 buffer | Amylase Starch pH 7.0 buffer | Amylase Deionized water pH 7.0 buffer | Deionized water Starch pH 7.0 buffer | Deionized water Maltose pH 7.0 buffer | Amylase Starch pH 2.0 buffer | Amylase Starch pH 9.0 buffer |
| **Incubation condition** | Boil first, then incubate at 37°C | 37°C | 37°C | 37°C | 37°C | 37°C | 37°C |
| **Benedict's test** | | | | | | | |
| **IKI test** | | | | | | | |

real life will take only 60 seconds in the simulation. When the incubation time elapses, the test tube rack will automatically rise, and the doors to the assay cabinet will open.

## Amylase Assays

After the assay cabinet doors open, notice the two reagents in the assay cabinet. IKI tests for the presence of starch and Benedict's detects the presence of maltose, the digestion product of starch. Below the reagents are seven small assay tubes into which you will dispense a small amount of test solution from the incubated samples in the incubation unit, plus a drop of IKI.

1. Click and hold the mouse on the first tube in the incubation unit. Notice that the mouse pointer is now a miniature test tube tilted to the left.

2. While still holding the mouse button down, move the mouse pointer to the first small assay tube on the left side of the assay cabinet. Release the mouse button. Watch the first test tube automatically decant approximately half of its contents into the first assay tube on the left.

3. Repeat steps 1 and 2 for the remaining tubes in the incubation unit, moving to a fresh assay tube each time.

4. Next, click and hold the mouse on the IKI dropper cap and drag it to the first assay tube. Release the mouse button to dispense a drop of IKI into the first assay tube on the left. You will see IKI drip into the tube, which may cause a color change in the solution. A blue-black color indicates a **positive starch test.** If starch is not present, the mixture will look like diluted IKI, a **negative starch test.** Intermediate starch amounts result in a pale gray color.

5. Now dispense IKI into the remaining assay tubes. Record your results ($+$ for positive, $-$ for negative) in Chart 1.

6. Dispense Benedict's reagent into the remaining mixture in each tube in the incubation unit by dragging-and-dropping the Benedict's dropper cap to the top of each test tube.

7. After Benedict's reagent has been delivered to each tube in the incubation unit, click **Boil.** The entire tube rack will descend into the incubation unit and automatically boil the tube contents for a few moments.

8. When the rack of tubes rises, inspect the tubes for color change. A green-to-reddish color indicates that maltose is present; this is a **positive sugar test.** An orange-colored sample contains more maltose than a green sample. A reddish-brown color indicates even more maltose. A negative sugar test is indicated by no color change from the original bright blue. Record your results in Chart 1.

9. Click **Record Data** to display your results in the grid and retain your data in the computer's memory for later analysis. To repeat the experiment, drag all test tubes to the test tube washer and start again.

10. Answer the following questions, referring to Chart 1 (or the data grid in the simulation) as necessary. Hint: closely examine the IKI and Benedict's results for each tube.

What do tubes 2, 6, and 7 reveal about pH and amylase activity?

_____

_____

Which pH buffer allowed the highest amylase activity?

_____

Which tube indicates that the amylase did not contain maltose? _____

Which tubes indicate that the deionized water did not contain starch or maltose? _____

If we left out control tubes 3, 4, or 5, what objections could be raised to the statement: "Amylase digests starch to maltose"? (Hint: think about the purity of the chemical solutions.)

_____

_____

Would the amylase present in saliva be active in the stomach? Explain your answer.

_____

_____

What effect does boiling have on enzyme activity?

_____

_____ ∎

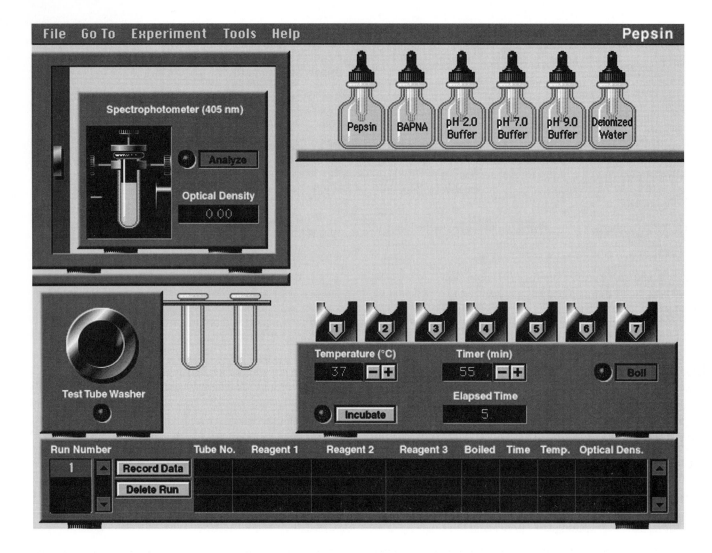

**Figure 40B.2 Opening screen of the Pepsin experiment.**

## Protein Digestion by Pepsin

The chief cells of the stomach glands produce pepsin, a protein-digesting enzyme. Pepsin hydrolyzes proteins to small fragments (proteoses, peptones, and peptides). In this experiment, you will use BAPNA, a synthetic "protein" that is transparent and colorless when in solution. However, if an active, protein-digesting enzyme such as pepsin is present, the solution will become yellow. You can use this characteristic to detect pepsin activity: the solution turns yellow if the enzyme digests the BAPNA substrate; it remains colorless if pepsin is not active or not present. One advantage of using a synthetic substrate is that you do not need any additional indicator reagents to see enzyme activity.

Choose **Pepsin** from the **Experiment** menu. The opening screen will appear in a few seconds (Figure 40B.2). The solutions shelf, test tube washer, and incubation equipment are the same as in the amylase experiment; only the solutions have changed.

Data displayed in the grid include the tube number, the three substances dispensed into each tube, a (+) or (−) mark indicating whether or not a sample was boiled, the time and temperature of the incubation, and the optical density measurement indicating enzyme assay results.

Activity:
# Assessing Protein Digestion by Pepsin

## Pepsin Incubation

1. Individually drag six test tubes to the test tube holders in the incubation unit.

2. Prepare the tubes with the substances indicated in Chart 2 using the following method.

- Click and hold the mouse button on the dropper cap of the desired substance and drag the dropper cap to the top of the desired test tube.

- Release the mouse button to dispense the substance.

3. Once all tubes are prepared, click the number (1) under the first test tube. The tube will descend into the incubation unit. All other tubes should remain in the raised position.

4. Click **Boil** to boil the number 1 tube. After boiling for a few moments, the tube will automatically rise.

5. Adjust the incubation temperature to 37°C and the timer to 60 min (compressed time) by clicking the (+) or (−) buttons.

6. Click **Incubate** to start the run. The incubation unit will gently agitate the test tube rack, evenly mixing the contents of all test tubes throughout the incubation. The computer is compressing the 60-minute time period into 60 seconds of real time. When the incubation time elapses, the test tube rack will automatically rise, and the doors to the assay cabinet will open.

## Pepsin Assay

After the assay cabinet doors open, you will see an instrument called a spectrophotometer, which you will use to measure how much yellow dye was liberated by pepsin digestion of BAPNA. When a test tube is dragged to the holder in the spectrophotometer and the **Analyze** button is clicked, the instrument will shine a light through a specimen to measure the amount of light absorbed by the sample within the tube. The measure of the amount of light absorbed by the solution is known as its *optical density*. A colorless solution does not absorb light, whereas a colored solution has a relatively high light absorbance. For example, a colorless solution has an optical density of 0.0. A colored solution, however, absorbs some of the light emitted by the spectrophotometer, resulting in an optical density reading greater than zero.

In this experiment a yellow-colored solution is a direct indication of the amount of BAPNA digested by pepsin. Although you can visually estimate the yellow color produced by pepsin digestion of BAPNA, the spectrophotometer precisely measures how much BAPNA digestion occurred in the experiment.

1. Click and hold the mouse on the first tube in the incubation unit and drag it to the holder in the spectrophotometer.

2. Release the mouse button to drop the tube into the holder.

3. Click **Analyze.** You will see light shining through the solution in the test tube as the spectrophotometer measures its optical density. The optical density of the sample will be displayed in the optical density window below the Analyze button.

| **Chart 2** | Pepsin Digestion of Protein | | | | | |
|---|---|---|---|---|---|---|
| **Tube no.** | **1** | **2** | **3** | **4** | **5** | **6** |
| **Additives** | Pepsin BAPNA pH 2.0 buffer | Pepsin BAPNA pH 2.0 buffer | Pepsin Deionized water pH 2.0 buffer | Deionized water BAPNA pH 2.0 buffer | Pepsin BAPNA pH 7.0 buffer | Pepsin BAPNA pH 9.0 buffer |
| **Incubation condition** | Boil first, then incubate at 37°C | 37°C | 37°C | 37°C | 37°C | 37°C |
| **Optical density** | | | | | | |

4. Record the optical density in Chart 2.

5. Drag the tube to its original position in the incubation unit and release the mouse button.

6. Repeat steps 1 through 5 for the remaining test tubes in the incubation unit.

7. Click **Record Data** to display your results in the grid and retain your data in the computer's memory for later analysis. To repeat the experiment, you must drag all test tubes to the test tube washer and start again.

8. Answer the following questions, referring to Chart 2 (or the data grid in the simulation) as necessary.

Which pH provided the highest pepsin activity? _____

Would pepsin be active in the mouth? Explain your answer.

_____

_____

How did the results of tube 1 compare with those of tube 2?

_____

Tubes 1 and 2 contained the same substances. Explain why their optical density measurements were different.

_____

_____

If you had not run the tube 2 and 3 samples, what argument could be made against the statement "Pepsin digests BAPNA"?

_____

_____

What do you think would happen if you reduced the incubation time to 30 minutes? Use the simulation to help you answer this question if you are not sure.

_____

_____

What do you think would happen if you decreased the temperature of incubation to 10°C? Use the simulation to help you answer this question if you are not sure.

_____

_____ ■

## Fat Digestion by Pancreatic Lipase and the Action of Bile

The treatment that fats and oils undergo during digestion in the small intestine is a bit more complicated than that of carbohydrates or proteins. Fats and oils require pretreatment with bile to physically emulsify the fats. As a result, two sets of reactions must occur.

First:

$$\text{Fats/oils} \xrightarrow[\text{(emulsification)}]{\text{bile}} \text{minute fat/oil droplets}$$

Then:

$$\text{Fat/oil droplets} \xrightarrow{\text{lipase}} \text{monoglycerides and fatty acids}$$

Lipase hydrolyzes fats and oils to their component monoglycerides and two fatty acids. Occasionally lipase hydrolyzes fats and oils to glycerol and three fatty acids.

The fact that some of the end products of fat digestion (fatty acids) are organic acids that decrease the pH provides an easy way to recognize that digestion is ongoing or completed. You will be using a pH meter in the assay cabinet to record the drop in pH as the test tube contents become acid.

Choose **Lipase** from the **Experiment** menu. The opening screen will appear in a few seconds (Figure 40B.3). The solutions shelf, test tube washer, and incubation equipment are the same as in the previous two experiments; only the solutions have changed.

Data displayed in the grid include the tube number, the four reagents dispensed into each tube, a (+) or (−) mark indicating whether or not a sample was boiled, the time and temperature of the incubation, and the pH measurement indicating enzyme assay results.

### Activity:

## Assessing Fat Digestion by Pancreatic Lipase and the Action of Bile

### Lipase Incubation

1. Individually drag 6 test tubes to the test tube holders in the incubation unit.

2. Prepare the tubes with the solutions indicated in Chart 3 on page 56 by using the following method.

• Click and hold the mouse button on the dropper cap of the desired substance.

• While holding the mouse button down, drag the dropper cap to the top of the desired test tube.

• Release the mouse button to dispense the substance.

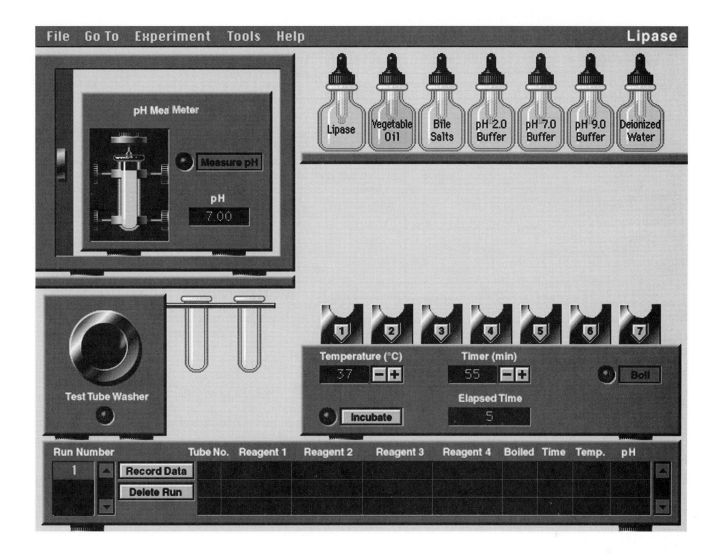

**Figure 40B.3    Opening screen of the Lipase experiment.**

3. Adjust the incubation temperature to 37°C and the timer to 60 min (compressed time) by clicking the (+) or (−) buttons.

4. Click **Incubate** to start the run. The incubation unit will gently agitate the test tube rack, evenly mixing the contents of all test tubes throughout the incubation. The computer is compressing the 60-minute time period into 60 seconds of real time. When the incubation time elapses, the test tube rack automatically rises, and the doors to the assay cabinet open.

## Lipase Assay

After the assay cabinet doors open, you will see a pH meter that you will use to measure the relative acidity of your test solutions. When a test tube is dragged to the holder in the pH meter and the Measure pH button is clicked, a probe will descend into the sample, take a pH reading, and then retract. The pH of the sample will be displayed in the pH window below the Measure pH button. A solution containing fatty acids liberated from fat by the action of lipase will exhibit a lower pH than one without fatty acids.

| Chart 3 | Pancreatic Lipase Digestion of Fats and the Action of Bile | | | | | |
|---|---|---|---|---|---|---|
| **Tube no.** | **1** | **2** | **3** | **4** | **5** | **6** |
| **Additives** | Lipase Vegetable oil Bile salts pH 7.0 buffer | Lipase Vegetable oil Deionized water pH 7.0 buffer | Lipase Deionized water Bile salts pH 9.0 buffer | Deionized water Vegetable oil Bile salts pH 7.0 buffer | Lipase Vegetable oil Bile salts pH 2.0 buffer | Lipase Vegetable oil Bile salts pH 9.0 buffer |
| **Incubation condition** | 37°C | 37°C | 37°C | 37°C | 37°C | 37°C |
| **pH** | | | | | | |

1.  Click and hold the mouse on the first tube in the incubation unit and drag it to the holder in the pH meter. Release the mouse button to drop the tube into the holder.

2.  Click **Measure pH.**

3.  In Chart 3, record the pH displayed in the pH window.

4.  Drag the test tube in the pH meter to its original position in the incubation unit and release the mouse button.

5.  Repeat steps 1 through 4 for the remaining test tubes in the incubation unit.

6.  Click **Record Data** to display your results in the grid and retain your data in the computer's memory for later analysis. To repeat the experiment, you must drag all test tubes to the test tube washer and begin again.

7.  Answer the following questions, referring to Chart 3 (or the data grid in the simulation) as necessary.

Explain the difference in activity between tubes 1 and 2.

_____

_____

Can we determine if fat hydrolysis has occurred in tube 6?

_____ Explain your answer. _____

_____

Which pH resulted in maximum lipase activity? _____

Is this method of assay sufficient to determine if the optimum

activity of lipase is at pH 2.0? _____

In theory, would lipase be active in the mouth? _____

 Would it be active in the stomach? _____

Explain your answers. _____

_____

Based on the enzyme pH optima you determined, where in the body would we expect to find the enzymes in these experiments?

Amylase _____

Pepsin _____

Lipase _____ ■

## Activity:
# Using Your Data

### Save Your Data

When you have finished with all experiments, you can save your data set for later review and additional analysis.

1. Select **Save Data to File** from the **File** menu.

2. When the dialog box opens, click a drive and/or directory location.

3. Type a file name into the **File name** box (your instructor may provide you with a name).

4. Click **Save** to complete the save. If you want to review previously saved data, select **Open Data File** from the **File** menu, and choose your file.

### Print Your Data

If your instructor requires a hard copy of your experiment results, choose **Print Data** from the **File** menu.

1. Type your name into the provided space.

2. When you are ready to print, click **OK.** ■

# Physical Processes: Mechanisms of Food Propulsion and Mixing

Although enzyme activity is an essential part of the overall digestion process, food must also be processed physically by churning and chewing, and moved by mechanical means along the tract if digestion and absorption are to be completed. Just about any time organs exhibit mobility, muscles are involved, and movements of and in the gastrointestinal tract are no exception. Although we tend to think only of smooth muscles for visceral activities, both skeletal and smooth muscles are necessary in digestion. This fact is demonstrated by the simple activities that follow.

## Activity:
# Studying Mechanisms of Food Propulsion and Mixing

### Deglutition (Swallowing)

*Swallowing,* or *deglutition,* which is largely the result of skeletal muscle activity, occurs in two phases: *buccal* (mouth) and *pharyngeal-esophageal.* The initial phase—the buccal—is voluntarily controlled and initiated by the tongue. Once begun, the process continues involuntarily in the pharynx and esophagus, through peristalsis, resulting in the delivery of the swallowed contents to the stomach.

Obtain a pitcher of water, a stethoscope, a paper cup, alcohol swabs, and an autoclave bag in preparation for making the following observations.

1. While swallowing a mouthful of water, consciously note the movement of your tongue during the process. Record your observations.

_____

_____

_____

2. Repeat the swallowing process while your laboratory partner watches the externally visible movements of your larynx. This movement is more obvious in a male, who has a larger Adam's apple. Record your observations.

_____

_____

What do these movements accomplish? _____

_____

_____

3. Your lab partner should clean the ear pieces of a stethoscope with an alcohol swab and don the stethoscope. Then your lab partner should place the diaphragm of the stethoscope on your abdominal wall, approximately 1 inch below the xiphoid process and slightly to the left, to listen for sounds as you again take two or three swallows of water. There should be two audible sounds. The first sound occurs when the water splashes against the gastroesophageal sphincter. The second occurs when the peristaltic wave of the esophagus arrives at the sphincter and the sphincter opens, allowing water to gurgle into the stomach. Determine, as accurately as possible, the time interval between these two sounds and record it below.

Interval between arrival of water at the sphincter and the opening of the sphincter:

_____ sec.

This interval gives a fair indication of the time it takes for the peristaltic wave to travel down the 10-inch-long esophagus. Actually the time interval is slightly less than it seems because pressure causes the sphincter to relax before the peristaltic wave reaches it.

Dispose of the used paper cup in the autoclave bag. ■

## Segmentation and Peristalsis

Although several types of movement occur in the digestive tract organs, segmentation and peristalsis are most important as mixing and propulsive mechanisms.

Segmental movements are local constrictions of the organ wall that occur rhythmically. They serve mainly to mix the foodstuffs with digestive juices and to increase the rate of absorption by continually moving different portions of the chyme over adjacent regions of the intestinal wall. However, segmentation is an important means of food propulsion in the small intestine, and slow segmenting movements called haustral contractions are common in the large intestine.

Peristaltic movements are the major means of propelling food through most of the digestive viscera. Essentially they are waves of contraction followed by waves of relaxation that squeeze foodstuffs through the alimentary canal, and they are superimposed on segmental movements.

### Activity:
## Viewing Propulsive Movements

If a videotape showing some of the propulsive movements is available, go to a viewing station to view it before leaving the laboratory. ■

# Respiratory System Mechanics: Computer Simulation

## Objectives

1. To define *ventilation, inspiration, expiration, forced expiration, tidal volume, vital capacity, expiratory reserve volume, inspiratory reserve volume, residual volume, surfactant, minute respiratory volume, forced expiratory volume,* and *pneumothorax.*

2. To describe the role of muscles and volume changes in the mechanics of breathing.

3. To understand that the lungs do not contain muscle and that respirations are therefore caused by external forces.

4. To explore the effect of changing airway resistance on breathing.

5. To study the effect of surfactant on lung function.

6. To examine the factors that cause lung collapse.

7. To understand the effects of hyperventilation, rebreathing, and breath holding on the $CO_2$ level in the blood.

## Materials

Minimum equipment required:

Windows:

- ❑ 486/66 MHz or better recommended
- ❑ Windows 95/98 recommended
- ❑ SVGA display (256 colors at 640 × 480)
- ❑ Minimum 16 MB available RAM
- ❑ Double-speed CD-ROM drive (quad-speed or higher recommended)
- ❑ Sound card
- ❑ Speakers or headphones
- ❑ Printer

Macintosh:

- ❑ 68040 processor or Power Macintosh recommended
- ❑ System 7.1 or higher
- ❑ Minimum 16 MB available RAM
- ❑ 13-inch or larger color monitor (640 × 480 resolution)

- ❑ Double-speed CD-ROM drive (quad-speed or higher recommended)
- ❑ Printer

Software:

- ❑ Benjamin/Cummings PhysioEx CD-ROM—Respiratory System Mechanics module

The two phases of **pulmonary ventilation** or **breathing** are **inspiration,** during which air is taken into the lungs, and **expiration,** during which air is expelled from the lungs. Inspiration occurs as the external intercostal muscles and the diaphragm contract. The diaphragm, normally a dome-shaped muscle, flattens as it moves inferiorly while the external intercostal muscles between the ribs lift the rib cage. These cooperative actions increase the thoracic volume. Because the increase in thoracic volume causes a partial vacuum, air rushes into the lungs. During normal expiration, the inspiratory muscles relax, causing the diaphragm to rise and the chest wall to move inward. The thorax returns to its normal shape due to the elastic properties of the lung and thoracic wall. Like a deflating balloon, the pressure in the lungs rises, which forces air out of the lungs and airways. Although expiration is normally a passive process, abdominal wall muscles and the internal intercostal muscles can contract to force air from the lungs. Blowing up a balloon is an example where such **forced expiration** would occur.

## Getting Started

Begin by making sure you have the computer equipment and software listed in the Materials section on this page.

1. Quit all applications currently running on your computer.

2. Insert the PhysioEx CD-ROM into the CD-ROM drive and keep PhysioEx in the drive during the entire time you use the program.

3. Follow the instructions below for your computer type.

**Windows 95/Windows 98**

The program will launch automatically when you load the CD-ROM into the drive. If autorun is disabled on your machine follow these instructions:

1. Double-click the **My Computer** icon on the Windows desktop.

2. Double-click the **PhysioEx** CD icon.

3. Look for the icon that looks like a tiny movie projector and double-click it.

**Macintosh**

Double-click the **PhysioEx** CD icon that appears in the window on your desktop to launch the program.

# Simulating Spirometry: Measuring Respiratory Volumes and Capacities

This computerized simulation allows you to investigate the basic mechanical function of the respiratory system as you determine lung volumes and capacities. The concepts you will learn by studying this simulated mechanical lung can then be applied to help you understand the operation of the human respiratory system.

Normal quiet breathing moves about 500 ml (0.5 liter) of air (the tidal volume) in and out of the lungs with each breath, but this amount can vary due to a person's size, sex, age, physical condition, and immediate respiratory needs. The terms used for the normal respiratory volumes are defined next. The values are for the normal adult male and are approximate.

## Normal Respiratory Volumes

**Tidal volume (TV):** Amount of air inhaled or exhaled with each breath under resting conditions (500 ml)

**Expiratory reserve volume (ERV):** Amount of air that can be forcefully exhaled after a normal tidal volume exhalation (1200 ml)

**Inspiratory reserve volume (IRV):** Amount of air that can be forcefully inhaled after a normal tidal volume inhalation (3100 ml)

**Residual volume (RV):** Amount of air remaining in the lungs after complete exhalation (1200 ml)

**Vital capacity (VC):** Maximum amount of air that can be exhaled after a normal maximal inspiration (4800 ml)

$$VC = TV + IRV + ERV$$

**Total lung capacity (TLC):** Sum of vital capacity and residual volume

## Pulmonary Function Tests

**Forced vital capacity (FVC):** Amount of air that can be expelled when the subject takes the deepest possible breath and exhales as completely and rapidly as possible

**Forced expiratory volume (FEV$_1$):** Measures the percentage of the vital capacity that is exhaled during 1 second of the FVC test (normally 75% to 85% of the vital capacity)

Choose **Respiratory System Mechanics** from the main menu. The opening screen for the Respiratory Volumes experiment will appear in a few seconds (Figure 48B.1). The main features on the screen when the program starts are a pair of *simulated lungs within a bell jar* at the left side of the screen, an *oscilloscope* at the upper right part of the screen, a *data display* area beneath the oscilloscope, and a *data control unit* at the bottom of the screen.

The black rubber "diaphragm" sealing the bottom of the glass bell jar is attached to a rod in the pump just below the jar. The rod moves the rubber diaphragm up and down to change the pressure within the bell jar (comparable to the intrapleural pressure in the body). As the diaphragm moves inferiorly, the resulting volume increase creates a partial vacuum in the bell jar because of lowered pressure. This partial vacuum causes air to be sucked into the tube at the top of the bell jar and then into the simulated lungs.

Conversely, as the diaphragm moves up, the rising pressure within the bell jar forces air out of the lungs. The partition between the two lungs compartmentalizes the bell jar into right and left sides. The lungs are connected to an airflow tube in which the diameter is adjustable by clicking the (+) and (−) buttons next to the Radius window in the equipment atop the bell jar. The volume of each breath that passes through the single air-flow tube above the bell jar is displayed in the Flow window. Clicking **Start** below the bell jar begins a trial run in which the simulated lungs will "breathe" in normal tidal volumes and the oscilloscope will display the tidal tracing. When **ERV** is clicked, the lungs will exhale maximally at the bottom of a tidal stroke and the expiratory reserve volume will be displayed in the Exp. Res. Vol. window below the oscilloscope. When **FVC** is clicked, the lungs will first inhale maximally and then exhale fully to demonstrate forced vital capacity. After ERV and FVC have been measured, the remaining lung values will be calculated and displayed in the small windows below the oscilloscope.

The data control equipment in the lower part of the screen records and displays data accumulated during the experiments. When you click **Record Data,** your data is recorded in the computer's memory and is displayed in the data grid. Data displayed in the data grid include the Radius, Flow, TV (tidal volume), ERV (expiratory reserve volume), IRV (inspiratory reserve volume), RV (residual volume), VC (vital capacity), FEV$_1$ (forced expiratory volume—1 second), TLC (total lung capacity), and Pump Rate. Clicking **Delete Line** allows you to discard the data for a single run; clicking **Clear Table** erases the entire experiment to allow you to start over.

If you need help identifying any piece of equipment, choose **Balloons On** from the Help menu and move the mouse pointer onto any piece of equipment visible on the computer's screen. As the pointer touches the object, a pop-up window appears to identify the equipment. To close the pop-up window, move the mouse pointer away from the equipment. Choose **Balloons On** again to turn off this help feature.

**A c t i v i t y :**

## Measuring Respiratory Volumes

Your first experiment will establish the baseline respiratory values.

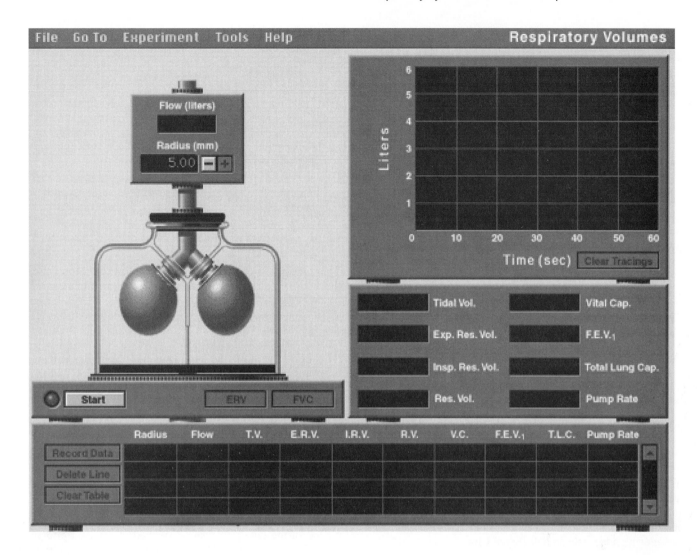

**Figure 48B.1   Opening screen of the Respiratory Volumes experiment.**

1.   If the grid in the data control unit is not empty, click **Clear Table** to discard all previous data.

2.   Adjust the radius of the airways to 5.00 mm by clicking the appropriate button next to the Radius window.

3.   Click **Start** and allow the tracing to complete. Watch the simulated lungs begin to breathe as a result of the "contraction and relaxation" of the diaphragm. Simultaneously, the oscilloscope will display a tracing of the tidal volume for each breath. The Flow window atop the bell jar indicates the tidal volume for each breath, and the Tidal Vol. window below the oscilloscope shows the average tidal volume. The Pump Rate window displays the number of breaths per minute.

4.   Click **Clear Tracings.**

5.   Now click **Start** again. After a second or two, click **ERV,** wait 2 seconds and then click **FVC** to complete the measurement of respiratory volumes. The expiratory reserve volume, inspiratory reserve volume, and residual volume will be automatically calculated and displayed from the tests you have performed so far. Also, the equipment calculates and displays the total lung capacity.

6.   Compute the **minute respiratory volume (MRV)** using the following formula (you can use the Calculator in the Tools menu):

$$MRV = TV \times BPM \text{ (breaths per minute)}$$

MRV _____ ml/min

7.   Does expiratory reserve volume include tidal volume?

_____

Explain your answer.

_____

_____

8.   Now click **Record Data** to record the current experimental data in the data grid. Then click **Clear Tracings.** ∎

### Activity:
## Examining the Effect of Changing Airway Resistance on Respiratory Volumes

Lung diseases are often classified as obstructive or restrictive. With an obstructive problem, expiratory flow is affected, whereas a restrictive problem might indicate reduced inspiratory volume. Although they are not diagnostic, pulmonary function tests such as $FEV_1$ can help a clinician determine the difference between obstructive and restrictive problems. $FEV_1$ is the forced volume exhaled in one second. In obstructive disorders like chronic bronchitis and asthma, airway resistance is increased and $FEV_1$ will be low. Here you will explore the effect of changing the diameter of the airway on pulmonary function.

1. Do *not* clear the data table from the previous experiment.

2. Adjust the radius of the airways to 4.50 mm by clicking the appropriate button next to the Radius window.

3. Click **Start** to begin respirations.

4. Click **FVC**. As you saw in the previous test, the simulated lungs will inhale maximally and then exhale as forcefully as possible. $FEV_1$ will be displayed in the $FEV_1$ window below the oscilloscope.

5. When the lungs stop respiring, click **Record Data** to record the current data in the data grid.

6. Decrease the radius of the airways in 0.50-mm decrements and repeat steps 4 and 5 until the minimum radius (3.00 mm) is achieved. Be sure to click **Record Data** after each trial. Click **Clear Tracings** between trials. If you make an error and want to delete a single value, click the data line in the data grid and then click **Delete Line.**

7. A useful way to express $FEV_1$ is as a percentage of the forced vital capacity. Copy the $FEV_1$ and vital capacity values from the computer screen to Chart 1, and then calculate the $FEV_1$ (%) by dividing the $FEV_1$ volume by the vital capacity volume. Record the $FEV_1$ (%) in Chart 1. You can use the Calculator under the Tools menu.

What happened to the $FEV_1$ (%) as the radius of the airways was decreased?

_____

Explain your answer.

_____

_____ ∎

## Simulating Factors Affecting Respirations

This part of the computer simulation allows you to explore the action of surfactant on pulmonary function and the effect of changing the intrapleural pressure.

Choose **Factors Affecting Respirations** from the **Experiment** menu. The opening screen will appear in a few seconds (Figure 48B.2). The basic features on the screen when the program starts are the same as in the Respiratory Volumes experiment screen. Additional equipment includes a surfactant dispenser atop the bell jar, and valves on each side of the bell jar. Each time **Surfactant** is clicked, a measured amount of surfactant is sprayed into the lungs. Clicking **Flush** washes surfactant from the lungs to prepare for another run.

Clicking **Open Valve** allows the pressure within that side of the bell jar to equalize with the atmospheric pressure. When **Reset** is clicked, the lungs are prepared for another run.

Data accumulated during a run are displayed in the windows below the oscilloscope. When you click **Record Data,** that data is recorded in the computer's memory and is displayed in the data grid. Data displayed in the data grid include the Radius, Pump Rate, the amount of Surfactant, Pressure Left (pressure in the left lung), Pressure Right (pressure in the right lung), Flow Left (air flow in the left lung), Flow Right (air flow in the right lung), and Total Flow. Clicking **Delete Line** allows you to discard data values for a single run, and clicking **Clear Table** erases the entire experiment to allow you to start over.

| **Chart 1** | FEV₁ as % of VC | | |
|---|---|---|---|
| **Radius** | **FEV₁** | **Vital Capacity** | **FEV₁ (%)** |
| 5.00 | | | |
| 4.50 | | | |
| 4.00 | | | |
| 3.50 | | | |
| 3.00 | | | |

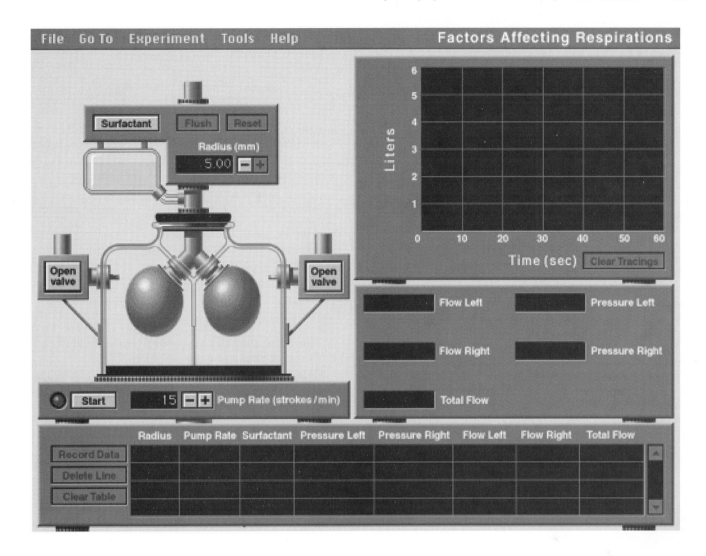

**Figure 48B.2    Opening screen of the Factors Affecting Respirations experiment.**

**Activity:**
## Examining the Effect of Surfactant

At any gas-liquid boundary, the molecules of the liquid are attracted more strongly to each other than they are to the air molecules. This unequal attraction produces tension at the liquid surface called *surface tension.* Because surface tension resists any force that tends to increase surface area, it acts to decrease the size of hollow spaces, such as the alveoli or microscopic air spaces within the lungs. If the film lining the air spaces in the lung were pure water, it would be very difficult, if not impossible, to inflate the lungs. However, the aqueous film covering the alveolar surfaces contains **surfactant,** a detergent-like lipoprotein that decreases surface tension by reducing the attraction of water molecules for each other. You will explore the action of surfactant in this experiment.

1.   If the data grid is not empty, click **Clear Table** to discard all previous data values.

2.   Adjust the airway radius to 5.00 mm by clicking the appropriate button next to the Radius window.

3.   If necessary, click **Flush** to clear the simulated lungs of existing surfactant.

4.   Click **Start** and allow a baseline run without added surfactant to complete.

5.   When the run completes, click **Record Data.**

6.   Now click **Surfactant** twice.

7.   Click **Start** to begin the surfactant run.

8.   When the lungs stop respiring, click **Record Data** to display the data in the grid.

How has the air flow changed compared to the baseline run?

_____

_____

Premature infants often have difficulty breathing. Explain why this might be so. (Use your text as needed.)

_____

_____ ■

## Activity:
# Investigating Intrapleural Pressure

The pressure within the pleural cavity, **intrapleural pressure,** is less than the pressure within the alveoli. This negative pressure condition is caused by two forces, the tendency of the lung to recoil due to its elastic properties and the surface tension of the alveolar fluid. These two forces act to pull the lungs away from the thoracic wall, creating a partial vacuum in the pleural cavity. Because the pressure in the intrapleural space is lower than atmospheric pressure, any opening created in the thoracic wall equalizes the intrapleural pressure with the atmospheric pressure, allowing air to enter the pleural cavity, a condition called **pneumothorax.** Pneumothorax allows lung collapse, a condition called **atelectasis** (at″ĕ-lik′tah-sis).

In the simulated respiratory system on the computer's screen, the intrapleural space is the space between the wall of the bell jar and the outer wall of the lung it contains. The pressure difference between inspiration and expiration for the left lung and for the right lung is individually displayed in the Pressure Left and Pressure Right windows.

1. Do *not* discard your previous data.

2. Click **Clear Tracings** to clean up the screen and then click **Flush** to clear the lungs of surfactant from the previous run.

3. Adjust the radius of the airways to 5.00 mm by clicking the appropriate button next to the Radius window.

4. Click **Start** and allow one screen of respirations to complete. Notice the negative pressure condition displayed below the oscilloscope when the lungs inhale.

5. When the lungs stop respiring, click **Record Data** to display the data in the grid.

6. Now click **Open Valve** on the left side of the bell jar above the Start button to open the valve.

7. Click **Start** to begin the run.

8. When the run completes, click **Record Data** again.

What happened to the lung in the left side of the bell jar?

_____

_____

How did the pressure in the left lung differ from that in the right lung?

_____

Explain your reasoning.

_____

_____

How did the total air flow in this trial compare with that in the previous trial in which the pleural cavities were intact?

_____

_____

What do you think would happen if the two lungs were in a single large cavity instead of separate cavities?

_____

_____

9. Now close the valve you opened earlier by clicking it and then click **Start** to begin a new trial.

10. When the run completes, click **Record Data** to display the data in the grid.

Did the deflated lung reinflate? _____
Explain your answer.

_____

_____

11. Click the **Reset** button atop the bell jar. This action draws the air out of the intrapleural space and returns it to normal resting condition.

12. Click **Start** and allow the run to complete.

13. When the run completes, click **Record Data** to display the data in the grid.

Why did lung function in the deflated (left) lung return to normal after you clicked Reset?

_____

_____ ■

# Simulating Variations in Breathing

This part of the computer simulation allows you to examine the effects of hyperventilation, rebreathing, and breathholding on $CO_2$ level in the blood.

Choose **Variations in Breathing** from the **Experiment** menu. The opening screen will appear in a few seconds (Figure 48B.3). The basic features on the screen when the pro-

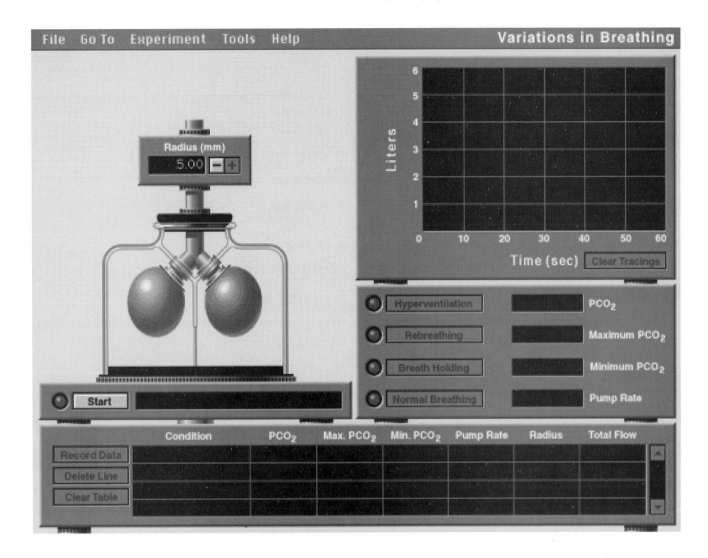

**Figure 48B.3    Opening screen of the Variations in Breathing experiment.**

gram starts are the same as in the lung volumes screen. The buttons beneath the oscilloscope control the various possible breathing patterns. Clicking **Hyperventilation** causes the lungs to breathe faster than normal. A small bag automatically covers the airway tube when **Rebreathing** is clicked. Clicking **Breath Holding** causes the lungs to stop respiring. Click **Normal Breathing** at any time to resume normal tidal cycles. The window next to the Start button displays the breathing pattern being performed by the simulated lungs.

The windows below the oscilloscope display the $PCO_2$ (partial pressure of $CO_2$) of the air in the lungs, Maximum $PCO_2$, Minimum $PCO_2$, and Pump rate.

Data accumulated during a run are displayed in the windows below the oscilloscope. When you click **Record Data,** that data is recorded in the computer's memory and is displayed in the data grid. Data displayed in the data grid include the Condition, $PCO_2$, Max. $PCO_2$, Min. $PCO_2$, Pump Rate, Radius, and Total Flow. Clicking **Delete Line** allows you to discard data values for a single run, and clicking **Clear Table** erases the entire experiment to allow you to start over.

A c t i v i t y :
## Exploring Various Breathing Patterns

You will establish the baseline respiratory values in this first experiment.

1.   If the grid in the data control unit is not empty, click **Clear Table** to discard all previous data.

2.   Adjust the radius of the airways to 5.00 mm by clicking the appropriate button next to the Radius window. Now, read through steps 3–5 before attempting to execute them.

3.   Click **Start** and notice that it changes to **Stop** to allow you to stop the respiration. Watch the simulated lungs begin to breathe as a result of the external mechanical forces supplied by the pump below the bell jar. Simultaneously, the oscilloscope will display a tracing of the tidal volume for each breath.

4.   After 2 seconds, click the **Hyperventilation** button and watch the $PCO_2$ displays. The breathing pattern will change to short, rapid breaths. The $PCO_2$ of the air in the lungs will be displayed in the small window to the right of the Hyperventilation button.

5. Watch the oscilloscope display and the PCO$_2$ window, and click **Stop** before the tracing reaches the end of the screen.

What happens to PCO$_2$ during hyperventilation? Explain your answer.

_____

_____

_____

6. Click **Record Data.**

7. Now click **Clear Tracings** to prepare for the next run.

## Rebreathing

When Rebreathing is clicked, a small bag will appear over the end of the air tube to allow the air within the lungs to be repeatedly inspired and expired.

1. Click **Start,** wait 2 seconds, and then click **Rebreathing.**

2. Watch the breathing pattern on the oscilloscope and notice the PCO$_2$ during the course of the run. Click **Stop** when the tracing reaches the right edge of the oscilloscope.

What happens to PCO$_2$ during the entire time of the rebreathing activity?

_____

_____

Did the depth of the breathing pattern change during rebreathing? (Carefully examine the tracing for rate and depth changes; the changes can be subtle.) Explain.

_____

_____

_____

_____

3. Click **Record Data** and then click **Clear Tracings** to prepare for the next run.

## Breath Holding

Breath holding can be considered an extreme form of rebreathing in which there is no gas exchange between the outside atmosphere and the air within the lungs.

1. Click **Start,** wait a second or two, and then click **Breath Holding.**

2. Let the breath-holding activity continue for about 5 seconds and then click **Normal Breathing.**

3. Click **Stop** when the tracing reaches the right edge of the oscilloscope.

What happened to the PCO$_2$ during breath holding?

_____

_____

What happened to the breathing pattern when normal respirations resume?

_____

_____

_____

4. Click **Record Data.** ■

## Activity:
# Using Your Data

## Save Your Data

When you have finished all parts of this exercise, you can save your data sets for later review and additional analysis.

1. Select **Save Data to File** from the **File** menu.

2. When the dialog box opens, click a drive and/or directory location.

3. Type a file name into the **File name** box (your instructor may provide you with a name).

4. Click **Save** to complete the save. If you want to review previously saved data, select **Open Data File** from the **File** menu, and choose your file.

## Print Your Data

If your instructor requires a printed copy of your experiment results, choose **Print** from the **File** menu.

1. Type your name into the space provided.

2. When you are ready to print, click **OK.** ■

# Renal Physiology—The Function of the Nephron: Computer Simulation

exercise
# 49B

## Objectives

**1.** To define *glomerulus, glomerular capsule, renal corpuscle, renal tubule, nephron, proximal convoluted tubule, loop of Henle,* and *distal convoluted tubule.*

**2.** To describe the blood supply to each nephron.

**3.** To identify the regions of the nephron involved in glomerular filtration and tubular reabsorption.

**4.** To study the factors affecting glomerular filtration.

**5.** To explore the concept of carrier transport maximum.

**6.** To understand how the hormones aldosterone and ADH affect the function of the kidney.

**7.** To describe how the kidneys can produce urine that is four times more concentrated than the blood.

## Materials

Minimum equipment required:

**Windows:**

❏ 486/66 MHz or better recommended

❏ Windows 95/98 recommended

❏ SVGA display (256 colors at 640 × 480)

❏ Minimum 16 MB available RAM

❏ Double-speed CD-ROM drive (quad-speed or higher recommended)

❏ Sound card

❏ Speakers or headphones

❏ Printer

**Macintosh:**

❏ 68040 processor or Power Macintosh recommended

❏ System 7.1 or higher

❏ Minimum 16 MB available RAM

❏ 13-inch or larger color monitor (640 × 480 resolution)

❏ Double speed CD-ROM drive (quad-speed or higher recommended)

❏ Printer

**Software:**

❏ Benjamin/Cummings PhysioEx CD-ROM—Renal System Physiology module

Metabolism produces wastes that must be eliminated from the body. This excretory function is the job of the renal system, most importantly the paired kidneys. Each kidney consists of about one million nephrons that carry out two crucial services, blood filtration and fluid processing.

## Microscopic Structure and Function of the Kidney

Each of the million or so **nephrons** in each kidney is a microscopic tubule consisting of two major parts: a **glomerulus** and a **renal tubule.** The glomerulus is a tangled capillary knot that filters fluid from the blood into the lumen of the renal tubule. The function of the renal tubule is to process that fluid, also called the **filtrate.** The beginning of the renal tubule is an enlarged end called the **glomerular capsule,** which surrounds the glomerulus and serves to funnel the filtrate into the rest of the renal tubule. Collectively, the glomerulus and the glomerular capsule are called the **renal corpuscle.**

As the rest of the renal tubule extends from the glomerular capsule, it becomes twisted and convoluted, then dips sharply down to form a hairpin loop, and then coils again before entering a collecting duct. Starting at the glomerular capsule, the anatomical parts of the renal tubule are as follows: the **proximal convoluted tubule,** the **loop of Henle** (nephron loop), and the **distal convoluted tubule.**

Two arterioles supply each glomerulus: an afferent arteriole feeds the glomerular capillary bed and an efferent arteriole drains it. These arterioles are responsible for blood flow through the glomerulus. Constricting the afferent arteriole lowers the downstream pressure in the glomerulus, whereas constricting the efferent arteriole will increase the pressure in the glomerulus. In addition, the diameter of the efferent arteriole is smaller than the diameter of the afferent arteriole, restricting blood flow out of the glomerulus. Consequently, the pressure in the glomerulus forces fluid through the endothelium of the glomerulus into the lumen of the surrounding glomerular capsule. In essence, everything in the blood except the cells and proteins are filtered through the glomerular wall. From the capsule, the filtrate moves into the rest of the renal tubule for processing. The job of the tubule is to reabsorb all the beneficial substances from its lumen while allowing the wastes to travel down the tubule for elimination from the body.

The nephron performs three important functions to process the filtrate into urine: glomerular filtration, tubular reabsorption, and tubular secretion. **Glomerular filtration** is a passive process in which fluid passes from the lumen of the glomerular capillary into the glomerular capsule of the renal tubule. **Tubular reabsorption** moves most of the filtrate back into the blood, leaving principally salt water plus the wastes in the lumen of the tubule. Some of the desirable or needed solutes are actively reabsorbed, and others move passively from the lumen of the tubule into the interstitial spaces. **Tubular secretion** is essentially the reverse of tubular reabsorption and is a process by which the kidneys can rid the blood of additional unwanted substances such as creatinine and ammonia.

The reabsorbed solutes and water that move into the interstitial space between the nephrons need to be returned to the blood, or the kidneys will rapidly swell like balloons. The peritubular capillaries surrounding the renal tubule reclaim the reabsorbed substances and return them to general circulation. Peritubular capillaries arise from the efferent arteriole exiting the glomerulus and empty into the veins leaving the kidney.

# Getting Started

Begin by making sure you have the computer equipment and software listed in the Materials section on p. 67.

1. Quit all applications currently running on your computer.

2. Insert the PhysioEx CD-ROM into the CD-ROM drive and keep PhysioEx in the drive during the entire time you use the program.

3. Follow the instructions below for your computer type.

**Windows 95/Windows 98**
The program will launch automatically when you load the CD-ROM into the drive. If autorun is disabled on your machine follow these instructions:

1. Double-click the **My Computer** icon on the Windows desktop.

2. Double-click the **PhysioEx CD** icon.

3. Look for the icon that looks like a tiny movie projector and double-click it.

**Macintosh**
Double-click the **PhysioEx** icon that appears in the window on your desktop to launch the program.

# Simulating Glomerular Filtration

This computerized simulation allows you to explore one function of a single simulated nephron, glomerular filtration. The concepts you will learn by studying a single nephron can then be applied to understand the function of the kidney as a whole.

Choose **Renal System Physiology** from the main menu. The opening screen for the Simulating Glomerular Filtration experiment will appear in a few seconds (Figure 49B.1). The main features on the screen when the program starts are a simulated blood supply at the left side of the screen, a simulated nephron within a supporting tank on the right side, and a data control unit at the bottom of the display.

The left beaker is the "blood" source representing the general circulation supplying the nephron. The "blood pressure" in the beaker is adjustable by clicking the (+) and (−) buttons on top of the beaker. A tube with an adjustable radius called the *afferent flow* tube connects the left beaker to the simulated glomerulus. Another adjustable tube called the *efferent flow* tube drains the glomerulus. The afferent flow tube represents the afferent arteriole feeding the glomerulus of each nephron, and the efferent flow tube represents the efferent arteriole draining the glomerulus. The outflow of the nephron empties into a collecting duct, which in turn drains into another small beaker at the bottom right part of the screen. Clicking the valve at the end of the collecting duct stops the flow of fluid through the nephron and collecting duct.

The Glomerular Pressure window on top of the nephron tank displays the pressure within the glomerulus. The Glomerular Filt. Rate window indicates the flow rate of the fluid moving from the lumen of the glomerulus into the lumen of the renal tubule.

The concentration gradient bathing the nephron is fixed at 1200 mosm. Clicking **Start** begins the experiment. Clicking **Refill** resets the equipment to begin another run.

The equipment in the lower part of the screen is called the *data control unit*. This equipment records and displays data you accumulate during the experiments. The data set for the first experiment (Afferent) is highlighted in the **Data Sets** window. You can add or delete a data set by clicking the appropriate button to the right of the Data Sets window. When you click **Record Data,** your data is recorded in the computer's memory and is displayed in the data grid. Data displayed in the data grid include the Afferent Radius, Efferent Radius, Beaker Pressure, Glomerular Pressure, Glomerular Filtration Rate, and the Urine Volume. Clicking **Delete Line** allows you to discard data values for a single run, and clicking **Clear Data Set** erases the entire experiment to allow you to start over.

If you need help identifying any piece of equipment, choose **Balloons On** from the Help menu and move the mouse pointer onto any piece of equipment visible on the computer's screen. As the pointer touches the object, a pop-up window appears identifying the equipment. To close the pop-up window, move the mouse pointer away from the equipment. Choose **Balloons On** again to turn off this help feature.

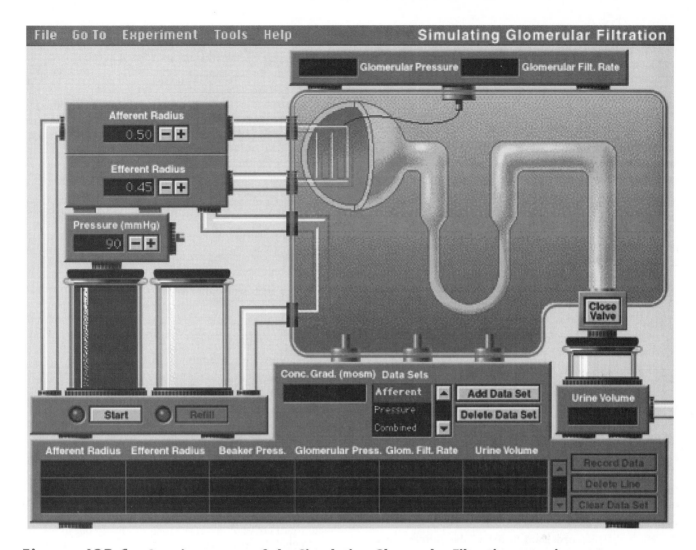

**Figure 49B.1    Opening screen of the Simulating Glomerular Filtration experiment.**

Activity:

# Investigating the Effect of Flow Tube Radius on Glomerular Filtration

Your first experiment will examine the effects of flow tube radii and pressures on the rate of glomerular filtration. Click **Start** to see the on-screen action. Continue when you understand how the simulation operates. Click **Refill** to reset the experiment.

1.   The **Afferent** line in the Data Sets window of the data control unit should be highlighted in bright blue. If it is not, choose it by clicking the **Afferent** line. The data control unit will now record filtration rate variations due to changing afferent flow tube radius.

2.   If the data grid is not empty, click **Clear Data Set** to discard all previous data.

3.   Adjust the afferent radius to 0.35 mm, and the efferent radius to 0.40 mm by clicking the appropriate buttons.

4.   If the left beaker is not full, click **Refill.**

5.   Keep the beaker pressure at 90 mm Hg during this part of the experiment.

6.   Click **Start** and watch the blood flow. Simultaneously, filtered fluid will be moving through the nephron and into the collecting duct. The Glomerular Filt. Rate window will display the fluid flow rate into the renal tubule when the left beaker has finished draining.

7.   Now click **Record Data** to record the current experiment data in the data grid. Click **Refill** to replenish the left beaker and prepare the nephron for the next run.

8.   Increase the afferent radius in 0.05-mm increments and repeat steps 6 through 8 until the maximum radius (0.60 mm) is achieved. Be sure to click **Record Data** after each trial. If you make an error and want to delete a single value, click the data line in the data grid and then click **Delete Line.**

What happens to the glomerular filtration rate as the afferent radius is increased?

_____

_____

Predict the effect of increasing or decreasing the efferent radius on glomerular filtration rate. Use the simulation to reach an answer if you are not sure.

_____

_____ ■

## Activity:
## Studying the Effect of Pressure on Glomerular Filtration

Both the blood pressure supplying the glomerulus and the pressure in the renal tubule have a significant impact on the glomerular filtration rate. In this activity, the data control unit will record filtration rate variations due to changing pressure.

1. Click the **Pressure** line in the Data Sets window of the data control unit.

2. If the data grid is not empty, click **Clear Data Set** to discard all previous data.

3. If the left beaker is not full, click **Refill.**

4. Adjust the pressure in the left beaker to 70 mm Hg by clicking the appropriate button.

5. During this part of the experiment, maintain the afferent flow tube radius at 0.55 mm and the efferent flow tube radius at 0.45 mm.

6. Click **Start** and watch the blood flow. Filtrate will move through the nephron into the collecting duct. At the end of the run, the Glomerular Filt. Rate window will display the filtrate flow rate into the renal tubule.

7. Now click **Record Data** to record the current experiment data in the data grid. Click **Refill** to replenish the left beaker.

8. Increase the pressure in the upper beaker in increments of 10 mm Hg and repeat steps 6 through 8 until the maximum pressure (100 mm Hg) is achieved. Be sure to click **Record Data** after each trial. If you make an error and want to delete a single value, click the data line in the data grid and then click **Delete Line.**

What happened to the glomerular filtration rate as the beaker pressure was increased?

_____

Explain your answer.

_____

_____

_____

_____ ■

## Activity:
## Assessing Combined Effects on Glomerular Filtration

So far, you have examined the effects of flow tube radius and pressure on glomerular filtration rate. In this experiment you will be altering both variables to explore the combined effects on glomerular filtration rate and to see how one can compensate for the other to maintain an adequate glomerular filtration rate.

1. Click **Combined** in the Data Sets window of the data control unit.

2. If the data grid is not empty, click **Clear Data Set** to discard all previous data.

3. If the left beaker is not full, click **Refill.**

4. Set the starting conditions at

• 100 mm Hg beaker pressure

• 0.55 mm afferent radius

• 0.45 mm efferent radius

5. Click **Start.**

6. Now click **Record Data** to record the current baseline data in the data grid.

7. Click **Refill.**

You will use this baseline data to compare a run in which the valve at the end of the collecting duct is in the open position with a run in which the valve is in the closed position.

8. Use the simulation and your knowledge of basic renal anatomy to arrive at answers to the following questions. Be sure to click **Record Data** after each trial. If you make an error and want to delete a single value, click the data line in the data grid and then click **Delete Line.**

Click **Close Valve** on the end of the collecting duct to close it. Click **Start** and allow the run to complete. How does this run compare to the runs in which the valve was open?

_____

Expanding on this concept, what might happen to total glomerular filtration and therefore urine production in a human kidney if all of its collecting ducts were totally blocked?

_____

_____

Would kidney function as a whole be affected if a single nephron was blocked? Explain.

_____

_____

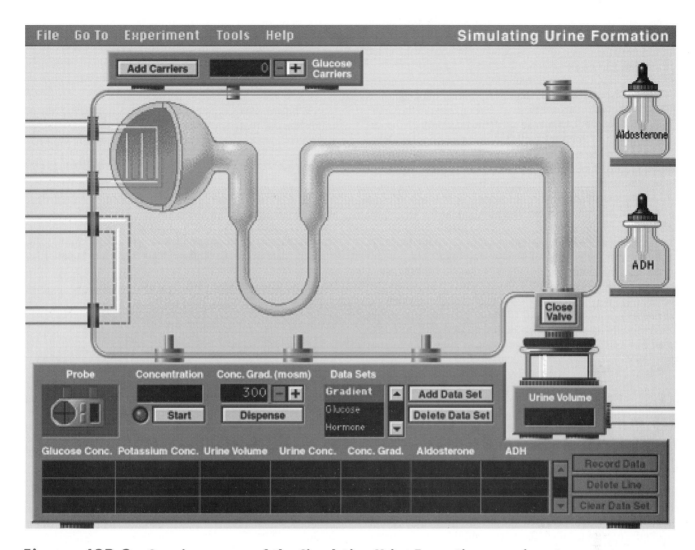

**Figure 48B.2    Opening screen of the Simulating Urine Formation experiment.**

Would the kidney be functioning if glomerular filtration was zero? Explain.

_____

_____

Explain how the body could increase glomerular filtration rate in a human kidney.

_____

_____

_____

_____

If you increased the pressure in the beaker, what other condi-

tion(s) could you adjust to keep the glomerular filtration rate constant?

_____

_____    ■

## Simulating Urine Formation

This part of the computer simulation allows you to explore some aspects of urine formation by manipulating the interstitial solute concentration. Other activities include investigating the effects of aldosterone and ADH (antidiuretic hormone), and the role that glucose carrier proteins play in renal function.

Choose **Simulating Urine Formation** from the **Experiment** menu. The opening screen will appear in a few seconds (Figure 49B.2). The basic features on the screen when the program starts are similar to the glomerular filtration screen. Most of the vascular controls have been moved off-screen to the left because they will not be needed in this set of experiments. Additional equipment includes a *supplies shelf* at the right side of the screen, a *glucose carrier control* located at the top of the nephron tank, and a concentration probe at the bottom left part of the screen.

The maximum concentration of the "interstitial gradient" to be dispensed into the tank surrounding the nephron is adjusted by clicking the (+) and (−) buttons next to the Conc. Grad. window. Click **Dispense** to fill the tank through the jets at the bottom of the tank with the chosen solute gradient. Click **Start** to begin a run. After a run completes, the concentration probe can be clicked and dragged over the nephron to display the solute concentration within.

Hormone is dispensed by dragging a hormone bottle cap to the gray cap button in the nephron tank at the top of the collecting duct and then letting go of the mouse button.

The (+) and (−) buttons in the glucose carrier control are used to adjust the number of glucose carriers that will be inserted into the simulated proximal convoluted tubule when the Add Carriers button is clicked.

Data displayed in the data grid will depend on which experiment is being conducted. Clicking **Delete Line** allows you to discard data values for a single run, and clicking **Clear Data Set** erases the entire experiment to allow you to start over.

## Activity:
# Exploring the Role of the Solute Gradient on Maximum Urine Concentration Achievable

In the process of urine formation, solutes and water move from the lumen of the nephron into the interstitial spaces. The passive movement of solutes and water from the lumen of the renal tubule into the interstitial spaces relies in part on the total solute gradient surrounding the nephron. When the nephron is permeable to solutes or water, an equilibrium will be reached between the interstitial fluid and the contents of the nephron. Antidiuretic hormone (ADH) increases the water permeability of the distal convoluted tubule and the collecting duct, allowing water to flow to areas of higher solute concentration, usually from the lumen of the nephron into the surrounding interstitial area. You will explore the process of passive reabsorption in this experiment. While doing this part of the simulation, assume that when ADH is present the conditions favor the formation of the most concentrated urine possible.

1. **Gradient** in the Data Sets window of the data control unit should be highlighted in bright blue. If it is not then click Gradient.

2. If the data grid is not empty, click **Clear Data Set** to discard all previous data.

3. Click and hold the mouse button on the ADH bottle cap and drag it to the gray cap at the top right side of the nephron tank. Release the mouse button to dispense ADH onto the collecting duct.

4. Adjust the maximum total solute concentration of the gradient (**Conc. Grad.**) to 300 mosm by clicking the appropriate button. Because the blood solute concentration is also 300 mosm, there is no osmotic difference between the lumen

of the nephron and the surrounding interstitial fluid.

5. Click **Dispense.**

6. Click **Start** to begin the experiment. Filtrate will move through the nephron and then drain into the beaker below the collecting duct.

7. While the experiment is running, click and hold the mouse on the Probe and drag it to the urine beaker. Observe the total solute concentration in the Concentration window.

8. Now click **Record Data** to record the current experiment data in the data grid.

9. Increase the maximum concentration of the gradient in 300-mosm increments and repeat steps 3 through 8 until 1200 mosm is achieved. Be sure to click **Record Data** after each trial. If you make an error and want to delete a single value, click the data line in the data grid and then click **Delete Line.**

What happened to the urine concentration as the gradient concentration was increased?

_____

_____

What factor limits the maximum possible urine concentration?

_____

_____

The solute concentration of the blood is about 300 mosm, and the highest interstitial solute concentration in a human kidney is about 1200 mosm. This means that the maximum urine solute concentration is about four times that of the blood. What would be the maximum possible urine concentration if the maximum interstitial solute concentration were 3000 mosm instead of 1200 mosm? Explain. (Use the simulation to arrive at an answer if you are not sure.)

_____

_____

_____

_____  ■

## Activity:
# Studying the Effect of Glucose Carrier Proteins on Glucose Reabsorption

Because carrier proteins are needed to move glucose from the lumen of the nephron into the interstitial spaces, there is a limit to the amount of glucose that can be reabsorbed. When all glucose carriers are bound with the glucose they are transporting, excess glucose is eliminated in urine. In this experiment, you will examine the effect of varying the number of glucose transport proteins in the proximal convoluted tubule.

1. Click **Glucose** in the Data Sets window of the data control unit.

2. If the data grid is not empty, click **Clear Data Set** to discard all previous data.

3. Set the concentration gradient (**Conc. Grad.**) to 1200 mosm.

4. Click **Dispense.**

5. Adjust the number of glucose carriers to 100 (an arbitrary figure) by clicking the appropriate button.

6. Click **Add Carriers.** This action inserts the specified number of glucose carrier proteins per unit area into the membrane of the proximal convoluted tubule.

7. Click **Start** to begin the run after the carriers have been added.

8. Click **Record Data** to record the current experiment data in the data grid. Glucose presence in the urine will be displayed in the data grid.

9. Now increase the number of glucose carrier proteins in the proximal convoluted tubule in increments of 100 glucose carriers and repeat steps 6 through 8 until the maximum number of glucose carrier proteins (500) is achieved. Be sure to click **Record Data** after each trial. If you make an error and want to delete a single value, click the data line in the data grid and then click **Delete Line.**

What happened to the amount of glucose present in the urine as the number of glucose carriers was increased?

_____

_____

The amount of glucose present in normal urine is minimal because there are normally enough glucose carriers present to handle the "traffic." Predict the consequence in the urine if there was more glucose than could be transported by the available number of glucose carrier proteins.

_____

_____

Explain why we would expect to find glucose in the urine of a diabetic person.

_____

_____ ■

## Activity:
# Testing the Effect of Hormones on Urine Formation

The concentration of the urine excreted by our kidneys changes depending on our immediate needs. For example, if a person consumes a large quantity of water, the excess water will be eliminated, producing dilute urine. On the other hand, under conditions of dehydration, there is a clear benefit in being able to produce urine as concentrated as possible, thereby retaining precious water. Although the medullary gradient makes it possible to excrete concentrated urine, urine dilution or concentration is ultimately under hormonal control. In this experiment, you will investigate the effects of two different hormones on renal function, aldosterone produced by the adrenal gland and ADH manufactured by the hypothalamus and stored in the posterior pituitary gland. Aldosterone works to reabsorb $Na^+$ (and thereby water) at the expense of losing $K^+$. Its site of action is the distal convoluted tubule. ADH makes the distal tubule and collecting duct more permeable to water, thereby allowing the body to reabsorb more water from the filtrate when it is present.

1. Click **Hormone** in the Data Sets window of the data control unit.

2. If the data grid is not empty, click **Clear Data Set** to discard all previous data.

3. During this part of the experiment, keep the concentration gradient at 1200 mosm.

4. Click **Dispense** to add the gradient and then click **Start** to begin the experiment.

5. Now click **Record Data** to record the current experiment data in the data grid.

You will use this baseline data to compare with the conditions of the filtrate under the control of the two hormones.

6. Keeping all experiment conditions the same as before, do the following:

• Drag the aldosterone bottle cap to the gray cap on the top right side of the nephron tank and release the mouse to automatically dispense aldosterone into the tank surrounding the distal convoluted tubule and collecting duct.

• Click **Start** and allow the run to complete.

• Click **Record Data.**

In this run, how does the volume of urine differ from the previously measured baseline volume?

_____

_____

Explain the difference in the total amount of potassium in the urine between this run and the baseline run.

_____

_____

7. Drag the ADH bottle cap to the gray cap on the top right side of the nephron tank and release it to dispense ADH.

• Click **Start** and allow the run to complete

• Click **Record Data.**

In this run, how does the volume of urine differ from the baseline measurement?

_____

_____

Is there a difference in the total amount of potassium in this run and the total amount of potassium in the baseline run? Explain. (Hint: the urine volume with ADH present is about one-tenth the urine volume when it is not present.)

_____

_____

Are the effects of aldosterone and ADH similar or antagonistic? _____

Consider this situation: we want to reabsorb sodium ions but do not want to increase the volume of the blood by reabsorbing water from the filtrate. Assuming that aldosterone and ADH are both present, how would you adjust the hormones to accomplish the task?

_____

_____

_____

_____

If the interstitial gradient ranged from 300 mosm to 3000 mosm and ADH was not present, what would be the maximum possible urine concentration? _____ ■

### A c t i v i t y :
## Using Your Data

### Save Your Data

When you have finished all parts of this exercise, you can save your data sets for later review and additional analysis.

1.  Select **Save Data to File** from the **File** menu.

2.  When the dialog box opens, click a drive and/or directory location.

3.  Type a file name into the **File name** box (your instructor may provide you with a name).

4.  Click **Save** to complete the save. If you want to review previously saved data, select **Open Data File** from the **File** menu, and choose your file.

### Print Your Data

If your instructor requires a printed copy of your experiment results, choose **Print** from the **File** menu.

1.  Type your name into the space provided.

2.  When you are ready to print, click **OK.** ■

# Using the Histology Module

Examining a specimen using a microscope accomplishes two goals: first, it gives you an understanding of the cellular organization of tissues, and second, perhaps even more importantly, it hones your observational skills. Because developing these skills is crucial in the understanding and eventual mastery of the way of thinking in science, using this histology module is not intended as a substitute for using the microscope. Instead, use the histology module to gain an overall appreciation of the specimens and then make your own observations using your microscope. The histology module is also an excellent review tool.

## Materials

Minimum equipment required:

**Windows:**

❑ 486/66 MHz or better recommended

❑ Windows 95/98 recommended

❑ SVGA display (256 colors at 640 × 480)

❑ Minimum 16 MB available RAM

❑ Double-speed CD-ROM drive (quad-speed or higher recommended)

**Macintosh:**

❑ 68040 processor or Power Macintosh recommended

❑ System 7.1 or higher

❑ Minimum 16 MB available RAM

❑ 13-inch or larger color monitor (640 × 480 resolution)

❑ Double-speed CD-ROM drive (quad-speed or higher recommended)

**Software:**

❑ Benjamin/Cummings PhysioEx CD-ROM—Histology module

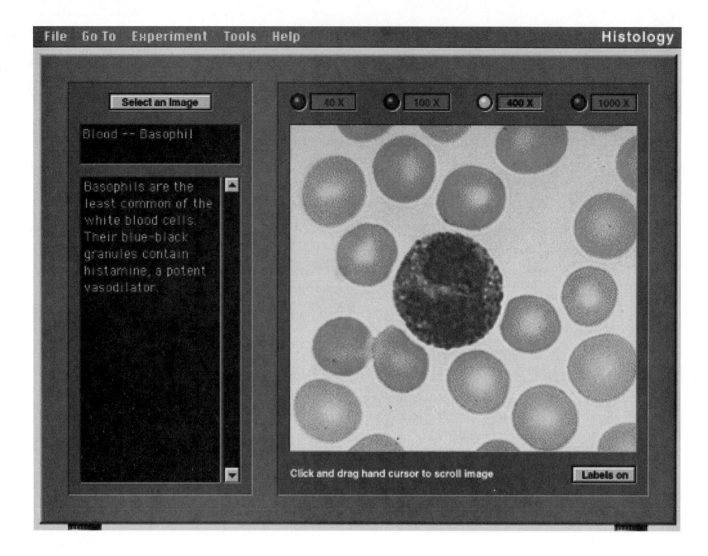

File  Go To  Experiment  Tools  Help                                    Histology

Select an Image

Blood -- Basophil

Basophils are the
least common of the
white blood cells.
Their blue-black
granules contain
histamine, a potent
vasodilator.

40 X    100 X    400 X    1000 X

Click and drag hand cursor to scroll image        Labels on

**Figure H    Opening screen of the Histology module.**

# Getting Started

1.   Quit all applications currently running on your computer.

2.   Insert the PhysioEx CD-ROM into the CD-ROM drive and keep PhysioEx in the drive during the entire time you use the program.

3.   Follow the instructions below for your computer type.

### Windows 95/Windows 98
The program will launch automatically when you load the CD-ROM into the drive. If autorun is disabled on your machine follow these instructions:

1.   Double-click the **My Computer** icon on the Windows desktop.

2.   Double-click the **PhysioEx** CD icon.

3.   Look for the icon that looks like a tiny movie projector and double-click it.

### Macintosh
Double-click the **PhysioEx** icon that appears in the window on your desktop to launch the program.

# Exploring Digital Histology

Choose **Histology** from the main menu. The opening screen for the Histology module will appear in a few seconds (Figure H). The main features on the screen when the program starts are a pair of empty text boxes at the left side of the screen, a large image-viewing window on the right side, and a set of magnification buttons above the image window.

Click **Select an Image** to bring up the image list. Choosing an image from the list displays it in the image-viewing window. The text description of the specimen is displayed in the large text box at the lower left and the image title in the small upper text box. Some of the words in the text are highlighted in color; clicking those words displays another image that you can compare with the main image.

Click and hold the mouse button (left button on a PC) on the image and then drag the mouse to move other areas of the specimen into the image viewing window, much like moving a slide on a microscope stage.

Choose the magnification of the image in the viewing window by clicking one of the magnification buttons above the image window.

To see parts of the slide labeled, click **Labels on**. Click the button again to remove the labels.

# Cell Transport Mechanisms and Permeability: Computer Simulation

Choose all answers that apply to items 1 and 2, and place their letters on the response blanks to the right.

**1.** Differential permeability: _____

    a.   is also called selective permeability

    b.   refers to the ability of the plasma membrane to select what passes through it

    c.   implies that all substances pass through membranes without hindrance

    d.   keeps wastes inside the cell and nutrients outside the cell

**2.** Passive transport includes: _____

    a.  osmosis      b.  simple diffusion      c.  phagocytosis      d.  pinocytosis      e.  facilitated diffusion

**3.** The following refer to the dialysis simulation.

Did the 20 MWCO membrane exclude any solute(s)? _____

Which solute(s) passed through the 100 MWCO membrane? _____

Which solute exhibited the highest diffusion rate through the 100 MWCO membrane? _____

Explain why this is so: _____

**4.** The following refer to the facilitated diffusion simulation.

Are substances able to travel against their concentration gradient? _____

Name two ways to increase the rate of glucose transport. _____

Did NaCl affect glucose transport? _____

Does NaCl require a transport protein for diffusion? _____

**5.** The following refer to the osmosis simulation.

Does osmosis require energy? _____

Is water excluded by any of the dialysis membranes? _____

Is osmotic pressure generated if solutes freely diffuse? _____

Explain how solute concentration affects osmotic pressure. _____

_____

**6.** The following refer to the filtration simulation.

What does the simulated filtration membrane represent in a living organism? _____

_____

What characteristic of a solute determines whether or not it passes through a filtration membrane?

_____

Would filtration occur if we equalized the pressure on both sides of a filtration membrane? _____

**7.** The following questions refer to the active transport simulation.

Does the presence of glucose carrier proteins affect $Na^+$ transport? _____

Can $Na^+$ be transported against its concentration gradient? _____

Are $Na^+$ and $K^+$ transported in the same direction? _____

The ratio of $Na^+$ to $K^+$ transport is _____ $Na^+$ transported out of the cell for every _____ $K^+$ transported into the cell.

**8.** What single characteristic of the semipermeable membranes used in the simple diffusion and filtration experiments determines which substances pass through them? _____

In addition to this characteristic, what other factors influence the passage of substances through living membranes?

_____

**9.** Assume the left beaker contains 4 m$M$ NaCl, 9 m$M$ glucose, and 10 m$M$ albumin. The right beaker contains 10 m$M$ NaCl, 10 m$M$ glucose, and 40 m$M$ albumin. Furthermore, the dialysis membrane is permeable to all substances except albumin. State whether the substance will move (a) to the right beaker, (b) to the left beaker, or (c) not move.

Glucose _____        Albumin _____

Water _____          NaCl _____

**10.** Assume you are conducting the experiment illustrated below. Both hydrochloric acid (HCl) with a molecular weight of about 36.5 and ammonium hydroxide (NH$_4$OH) with a molecular weight of 35 are volatile and easily enter the gaseous state. When they meet, the following reaction will occur:

$$HCl + NH_4OH \rightarrow H_2O + NH_4Cl$$

Ammonium chloride (NH$_4$Cl) will be deposited on the glass tubing as a smoky precipitate where the two gases meet. Predict which gas will diffuse more quickly and indicate to which end of the tube the smoky precipitate will be closer.

a.   The faster diffusing gas is _____.

b.   The precipitate forms closer to the _____ end.

Rubber stopper     Cotton wad with HCl          Cotton wad with NH$_4$OH

Support

11. When food is pickled for human consumption, as much water as possible is removed from the food. What method is used to achieve this dehydrating effect? _____

_____

12. What determines whether a transport process is active or passive? _____

_____

13. Characterize passive and active transport as fully as possible by choosing all the phrases that apply and inserting their letters on the answer blanks.

Passive transport _____        Active transport _____

a.  accounts for the movement of fats and respiratory gases through the plasma membrane
b.  explains solute pumping, phagocytosis, and pinocytosis
c.  includes osmosis, simple diffusion, and filtration
d.  may occur against concentration and/or electrical gradients
e.  uses hydrostatic pressure or molecular energy as the driving force
f.  moves ions, amino acids, and some sugars across the plasma membrane

14. For the osmometer demonstration, explain why the level of the water column rose during the laboratory session.

_____

_____

_____

15. Define the following:

*diffusion:* _____

_____

*osmosis:* _____

_____

*simple diffusion:* _____

_____

*filtration:* _____

_____

*active transport:* _____

_____

*phagocytosis:* _____

_____

*pinocytosis:* _____

_____

*facilitated diffusion:* _____

_____

# Skeletal Muscle Physiology: Computer Simulation

## Electrical Stimulation

1. Complete the following statements by filling in your answer on the lines provided below.

A motor unit consists of a __a__ and all the __b__ it innervates. If a single motor unit is stimulated, it will respond in a(n) __c__ fashion, whereas whole muscle contraction is a(n) __d__ response. In order for muscles to work in a practical sense, __e__ is the method used to produce a slow, steady increase in muscle force.

When we see the slightest evidence of force production on a tracing, the stimulus applied must have reached __f__.

The weakest stimulation that will elicit the strongest contraction that a muscle is capable of is called the __g__. That level of contraction is called the __h__.

When the __i__ of stimulation is so high that the muscle tracing shows fused peaks, __j__ has been achieved.

a. _____       f. _____

b. _____       g. _____

c. _____       h. _____

d. _____       i. _____

e. _____       j. _____

2. Name each phase of a typical muscle twitch and describe what is happening in each phase.

a. _____

_____

b. _____

_____

c. _____

_____

3. Explain how the PhysioEx experimental muscle stimulation differs from the *in vivo* stimulation via the nervous system. (Note that the graded muscle response following both stimulation methods is similar.)

_____

_____

_____

**4.** What are the two *experimental* ways in which mode of stimulation can affect the muscle force?

_____ and _____

Explain your answer.

_____

_____

_____

# Isometric Contraction

**1.** Identify the following conditions by choosing one of the key terms listed on the right.

Key:

_____ is generated by muscle tissue when it is being stretched

a.   Total force

_____ requires the input of energy

b.   Resting force

_____ is measured by recording instrumentation during contraction

c.   Active force

**2.** Circle the correct response in the parentheses for each statement.

An increase in resting length results in an (increase/decrease) in passive force.

The active force initially (increased/decreased) and then (increased/decreased) as the resting length was increased from minimum to maximum.

As the total force increased, the active force (increased/decreased).

**3.** Explain what happens to muscle force production at extremes of length (too short or too long). Hint: think about sarcomere structure.

Muscle too short: _____

_____

Muscle too long: _____

_____

# Isotonic Contraction

**1.** Assuming a fixed starting length, describe the effect resistance has on the initial velocity of shortening, and explain why it has this effect.

_____

_____

_____

**2.** A muscle has just been stimulated under conditions that will allow both isometric and isotonic contractions. Describe what is happening in terms of length and force.

Isometric: _____

_____

Isotonic: _____

_____

# Terms

Select the condition from column B that most correctly identifies the term in column A.

| Column A | Column B |
|---|---|
| _____ 1. muscle twitch | a. response is all-or-none |
| _____ 2. wave summation | b. affects the force a muscle can generate |
| _____ 3. multiple motor unit summation | c. a single contraction of intact muscle |
| _____ 4. resting length | d. recruitment |
| _____ 5. resistance | e. increasing force produced by increasing stimulus frequency |
| _____ 6. initial velocity of shortening | f. muscle length changing due to relaxation |
| _____ 7. isotonic shortening | g. caused by application of maximal stimulus |
| _____ 8. isotonic lengthening | h. weight |
| _____ 9. motor unit | i. exhibits graded response |
| _____ 10. whole muscle | j. high values with low resistance values |
| _____ 11. tetanus | k. changing muscle length due to active forces |
| _____ 12. maximal response | l. recording shows no evidence of muscle relaxation |

# Cardiovascular Dynamics: Computer Simulation

For numbers 1 and 2 below, choose all answers that apply and place their letters on the response blanks to the right of the statement.

1. The circulation of blood through the vascular system is influenced by _____.

    a. blood viscosity
    b. the length of blood vessels
    c. the driving pressure behind the blood
    d. the radius or diameter of blood vessels

2. Peripheral resistance depends on _____.

    a. blood viscosity
    b. blood pressure
    c. vessel length
    d. vessel radius

3. Complete the following statements.

    The volume of blood remaining in the heart after ventricular contraction is called the _____.

    Cardiac output is defined as _____.

    The amount of blood pumped by the heart in a single beat is called the _____ volume.

    The human heart is actually two individual pumps working in _____.

    Stroke volume is calculated by _____.

    If stroke volume decreased, heart rate would _____ in order to maintain blood flow.

4. How could the heart compensate to maintain proper blood flow for the following conditions?

    High peripheral resistance: _____

    A leaky atrioventricular valve: _____

    A constricted semilunar valve: _____

5. How does the size of the heart change under conditions of chronic high peripheral resistance?

    _____

6. The following questions refer to the Vessel Resistance experiment.

    How was the flow rate affected when the radius of the flow tube was increased? _____

    Which of the adjustable parameters had the greatest effect on fluid flow? _____

    How does vessel length affect fluid flow? _____

If you increased fluid viscosity, what parameter(s) could you adjust to keep fluid flow constant?

_____

Explain your answer. _____

_____

If the driving pressure in the left beaker was 100 mm Hg, how could you adjust the conditions of the experiment to completely stop fluid flow?

pletely stop fluid flow? _____

7. The following questions refer to the Pump Mechanics experiment.

What would happen if the right side of the heart pumped faster than the left side of the heart?

_____

_____

When you change the radius of the right flow tube in Pump Mechanics the resulting plot looks different than the radius plot in the Vessel Resistance experiment. How would the plot look if you changed the radius of both flow tubes in Pump Mechanics instead of just the right flow tube?

_____

_____

Why are valves needed in the Pump Mechanics equipment? _____

What happens to blood flow if peripheral resistance equals pump pressure? _____

Theoretically, what would happen to the pumping ability of the heart if the end systolic volume was equal to the end diastolic volume?

stolic volume? _____

8. Match the part in the simulation equipment to the analogous cardiac structure or physiological term listed in the key below.

Simulation equipment:

Key:

_____ 1. valve leading to the right beaker

a. pulmonary veins

_____ 2. valve leading to the pump

b. bicuspid valve

_____ 3. left flow tube

c. ventricular filling pressure

_____ 4. right flow tube

d. peripheral resistance

_____ 5. pump end volume

e. aortic valve

_____ 6. pump starting volume

f. end diastolic volume

_____ 7. pressure in the right beaker

g. aorta

_____ 8. pressure in the left beaker

h. end systolic volume

_____ 9. pump pressure

i. systolic pressure

**9.** Define the following terms:

*blood flow:* _____

_____

*peripheral resistance:* _____

_____

*viscosity:* _____

_____

*radius:* _____

_____

*end diastolic volume:* _____

_____

*systole:* _____

_____

*diastole:* _____

_____

# Frog Cardiovascular Physiology: Computer Simulation

## Special Electrical Properties of Cardiac Muscle: Automaticity and Rhythmicity

1. Define the following terms:

   *Automaticity* _____

   *Rhythmicity* _____

2. Explain the anatomical differences between frog and human hearts.

   _____

   _____

   _____

## Baseline Frog Heart Activity

1. Define the following terms:

   *Intrinsic heart control* _____

   *Extrinsic heart control* _____

2. Why is it necessary to keep the frog heart moistened with Ringer's solution? _____

   _____

## Refractory Period of Cardiac Muscle

1. Define *extrasystole* _____

   _____

2. Refer to the exercise to answer the following questions.

   What was the effect of stimulating the heart during ventricular contraction? _____

   During ventricular relaxation _____

   During the pause interval _____

   What does this information indicate about the refractory period of cardiac muscle?

   _____

   Can cardiac muscle be tetanized? _____ Why or why not? _____

# The Effect of Vagus Nerve Stimulation

1. What was the effect of vagal stimulation on heart rate? _____.

2. What is vagal escape? _____

3. Why is the vagal escape valuable in maintaining homeostasis? _____

_____

# Physical and Chemical Modifiers of Heart Rate

1. Describe the effect of thermal factors on the frog heart.

   Cold _____ Heat _____

2. Which of the following factors caused the same, or very similar, heart rate-reducing effects: epinephrine, atropine, pilocarpine, digitalis, potassium ions.

   _____

   _____

   Which of the factors listed above would reverse or antagonize vagal effects? _____

   _____

3. Did administering any the following produce any changes in force of contraction (shown by peaks of increasing or decreasing height)? If so, explain the mechanism.

   Epinephrine _____

   _____

   Calcium ions _____

   _____

4. Excessive amounts of each of the following ions would most likely interfere with normal heart activity. Explain the type of changes caused in each case.

   $K^+$ _____

   $Ca^{2+}$ _____

   $Na^+$ _____

5. Define the following:

   *Parasympathomimetic* _____

   *Ectopic pacemaker* _____

6. Explain how digitalis works. _____

_____

# Chemical and Physical Processes of Digestion

## Chemical Digestion of Foodstuffs: Enzymatic Action

1. Match the following definitions with the proper choices from the key.

   Key:    a. catalyst        b. control        c. enzyme        d. substrate

   _____ 1. increases the rate of a chemical reaction without becoming part of the product

   _____ 2. provides a standard of comparison for test results

   _____ 3. biological catalyst: protein in nature

   _____ 4. substance on which an enzyme works

2. Name three characteristics of enzymes. _____

   _____

3. Explain the following statement: The enzymes of the digestive system are classified as hydrolases.

   _____

   _____

4. Fill in the chart below with what you have learned about the various digestive system enzymes encountered in this exercise.

| Enzyme | Organ producing it | Site of action | Substrate(s) | Optimal pH |
|---|---|---|---|---|
| Salivary amylase | | | | |
| Pepsin | | | | |
| Lipase (pancreatic) | | | | |

5. Name the end products of digestion for the following types of foods.

   Proteins: _____    Carbohydrates: _____

   Fats: _____ and _____

6. You used several different indicators or tests in the laboratory to determine the presence or absence of certain substances. Choose the correct test or indicator from the key to correspond to the condition described below.

Key:    a.  IKI      b.  Benedict's solution      c.  pH meter      d.  BAPNA

_____ 1.  used to test for protein hydrolysis, which was indicated by a yellow color

_____ 2.  used to test for the presence of starch, which was indicated by a blue-black color

_____ 3.  used to test for the presence of fatty acids

_____ 4.  used to test for the presence of maltose, which was indicated by a blue to green (or to rust) color change

7. The three-dimensional structure of a functional protein is altered by intense heat or nonphysiological pH even though peptide bonds may not break. Such a change in protein structure is called denaturation, and denatured enzymes are not functional. Explain why.

_____

_____

8. What experimental conditions in the simulation resulted in denatured enzymes? _____

_____

9. Complete the mechanism of absorption section in the chart below for each of the substances listed. Use a check mark to indicate whether the absorption would result in the movement of a substance into the blood capillaries or the lymph capillaries (lacteals).

| Substance | Mechanism of absorption | Blood | Lymph |
|-----------|------------------------|-------|-------|
| Monosaccharides | | | |
| Fatty acids and glycerol | | | |
| Amino acids | | | |

10. Imagine that you have been chewing a piece of bread for 5 to 6 minutes. How would you expect its taste to change during

this time? _____

11. People on a strict diet to lose weight begin to metabolize stored fats at an accelerated rate. How could this condition affect

blood pH? _____

# Starch Digestion by Salivary Amylase

1.  What conclusions can you draw when an experimental sample gives both a positive starch test and a positive maltose test?

    _____

2.  Why was 37°C the optimal incubation temperature? _____

    _____

3.  Why did very little, if any, starch digestion occur in tube 1? _____

4.  Why did very little starch digestion occur in tubes 6 and 7? _____

5.  Imagine that you have told a group of your peers that amylase is capable of digesting starch to maltose. If you had not run the experiment in control tubes 3, 4, and 5, what objections to your statement could be raised?

    _____

    _____

# Protein Digestion by Pepsin

1.  Why is an indicator reagent such as IKI or Benedict's solution not necessary when using a substrate like BAPNA?

    _____

2.  Trypsin is a pancreatic hydrolase present in the small intestine during digestion. Would trypsin work well in the stomach? Explain your answer.

    _____

3.  How does the optical density of a solution containing BAPNA relate to enzyme activity? _____

    _____

4.  What happens to pepsin activity as it reaches the small intestine? _____

# Fat Digestion by Pancreatic Lipase and the Action of Bile

1.  Why does the pH of a fatty solution decrease as enzymatic hydrolysis increases? _____

    _____

2.  How does bile affect fat digestion? _____

3.  Why is it not possible to determine the activity of lipase in the pH 2.0 buffer using the pH meter assay method?

    _____

4.  Why is bile not considered an enzyme? _____

# Physical Processes: Mechanisms of Food Propulsion and Mixing

Complete the following statements. Write your answers in the numbered spaces below.

Swallowing, or __1__, occurs in two phases—the __2__ and __3__. One of these phases, the __4__ phase, is voluntary. During the voluntary phase, the __5__ is used to push the food into the back of the throat. During swallowing, the __6__ rises to ensure that its passageway is covered by the epiglottis so that the ingested substances do not enter the respiratory passageways. It is possible to swallow water while standing on your head because the water is carried along the esophagus involuntarily by the process of __7__. The pressure exerted by the foodstuffs on the __8__ sphincter causes it to open, allowing the food to enter the stomach.

The two major types of propulsive movements that occur in the small intestine are __9__ and __10__. One of these movements, __11__, acts to continually mix the foods and to increase the absorption rate by moving different parts of the chyme mass over the intestinal mucosa, but it has less of a role in moving foods along the digestive tract.

1. _____

2. _____

3. _____

4. _____

5. _____

6. _____

7. _____

8. _____

9. _____

10. _____

11. _____

# Respiratory System Mechanics: Computer Simulation

Define the following terms:

**1.** Ventilation _____

**2.** Inspiration _____

**3.** Expiration _____

## Measuring Respiratory Volumes

**1.** Write the respiratory volume term and the normal value that is described by the following statements:

Volume of air present in the lungs after a forceful expiration _____

Volume of air that can be expired forcefully after a normal expiration _____

Volume of air that is breathed in and out during a normal respiration _____

Volume of air that can be inspired forcefully after a normal inspiration _____

Volume of air corresponding to TV + IRV + ERV _____

**2.** Fill in the formula for minute respiratory volume:

_____

## Examining the Effect of Changing Airway Resistance on Respiratory Volumes

**1.** Even though pulmonary function tests are not diagnostic, they can help determine the difference between

_____ and _____ disorders.

**2.** Chronic bronchitis and asthma are examples of _____ disorders.

**3.** Describe $FEV_1$: _____

_____

**4.** Explain the difference between FVC and $FEV_1$: _____

_____

**5.** What effect would increasing airway resistance have on $FEV_1$? _____

# Examining the Effect of Surfactant

1. Explain the term *surface tension*._____

   _____

2. Surfactant is a detergent-like _____.

3. How does surfactant work?_____

4. What might happen to ventilation if the watery film lining the alveoli did not contain surfactant?

   _____

# Investigating Intrapleural Pressure

Complete the following statements.

The pressure within the pleural cavity, __1__, is __2__ than the pressure within the alveoli. This __3__ pressure condition is caused by two forces, the tendency of the lung to recoil due to its __4__ properties and the __5__ of the alveolar fluid. These two forces act to pull the lungs away from the thoracic wall, creating a partial __6__ in the pleural cavity. Because the pressure in the __7__ space is lower than __8__, any opening created in the thoracic wall equalizes the intrapleural pressure with the atmospheric pressure, allowing air to enter the pleural cavity, a condition called __9__. Pneumothorax allows __10__, a condition called __11__.

1. _____

2. _____

3. _____

4. _____

5. _____

6. _____

7. _____

8. _____

9. _____

10. _____

11. _____

12. Why is the intrapleural pressure negative rather than positive?_____

    _____

    _____

13. Would intrapleural pressure be positive or negative when blowing up a balloon? Explain your answer.

    _____

    _____

# Exploring Various Breathing Patterns

1. Match the term listed in column B with the descriptive phrase in column A. (There may be more than one correct answer.)

   **Column A**

   _____1. causes a drop in carbon dioxide concentration in the blood

   _____2. results in lower blood pH

   _____3. stimulates an increase respiratory rate

   _____4. results in a lower respiratory rate

   _____5. can be considered an extreme form of rebreathing

   _____6. causes a rise in blood carbon dioxide

   **Column B**

   a. rebreathing

   b. hyperventilation

   c. breath holding

2. Because carbon dioxide is the main stimulus for respirations, what would happen to respiratory drive if you held your

   breath?_____

   _____

# Renal Physiology—The Function of the Nephron: Computer Simulation

Define the following terms:

**1.** Glomerulus _____

**2.** Renal tubule _____

**3.** Glomerular capsule _____

**4.** Renal corpuscle _____

**5.** Afferent arteriole _____

**6.** Efferent arteriole _____

## Investigating the Effect of Flow Tube Radius on Glomerular Filtration

**1.** In terms of the blood supply to and from the glomerulus, explain why the glomerular capillary bed is unusual.

_____

_____

**2.** How would pressure in the glomerulus be affected by constricting the afferent arteriole? Explain your answer.

_____

_____

**3.** How would pressure in the glomerulus be affected by constricting the efferent arteriole? Explain your answer.

_____

_____

## Assessing Combined Effects on Glomerular Filtration

**1.** If systemic blood pressure started to rise, what could the arterioles of the glomerulus do to keep glomerular filtration rate

constant?_____

_____

**2.** One of the experiments you performed in the simulation was to close the valve at the end of the collecting duct. Is closing that valve more like constricting an afferent arteriole or more like a kidney stone? Explain your answer.

_____

_____

3. Constricting the efferent arteriole would have the same effect on glomerular filtration as (constricting/dilating) the afferent arteriole.

_____

_____

# Exploring the Role of the Solute Gradient on Maximum Urine Concentration Achievable

Complete the following statements.

In the process of urine formation, solutes and water move from the __1__ of the nephron into the __2__ spaces. The passive movement of solutes and water from the lumen of the renal tubule into the interstitial spaces relies in part on the __3__ surrounding the nephron. When the nephron is permeable to solutes or water, and __4__ will be reached between the interstitial fluid and the contents of the nephron. __5__ is a hormone that increases the water permeability of the __6__ and the collecting duct, allowing water to flow to areas of higher solute concentration, usually from the lumen of the nephron into the surrounding interstitial area. __7__ is hormone that causes __8__ reabsorption at the expense of __9__ loss into the lumen of the tubule.

1. _____

2. _____

3. _____

4. _____

5. _____

6. _____

7. _____

8. _____

9. _____

10. Would the passive movement of substances occur if the interstitial solute concentration was the same as the filtrate solute concentration? Explain your answer.

_____

_____

# Studying the Effect of Glucose Carrier Proteins on Glucose Reabsorption

1. In terms of the function of the nephron, explain why one might find glucose in the urine exiting the collecting duct.

_____

_____

2. Imagine this scenario: a person has the normal number of glucose carriers in the nephrons yet has glucose in the urine. What could be the cause of this condition? (Hint: think about filtration rate.)

_____

_____

# Testing the Effect of Hormones on Urine Formation

Complete the following statements.

The concentration of the __1__ excreted by our kidneys changes depending on our immediate needs. For example, if a person consumes a large quantity of water, the excess water will be eliminated, producing __2__ urine. On the other hand, under conditions of dehydration, there is a clear benefit in being able to produce urine as __3__ as possible, thereby retaining precious water. Although the medullary gradient makes it possible to excrete concentrated urine, urine dilution or concentration is ultimately under __4__ control. In this experiment, you will investigate the effects of two different hormones on renal function, aldosterone produced by the __5__ and ADH manufactured by the __6__ and stored in the __7__. Aldosterone works to reabsorb __8__ (and thereby water) at the expense of losing __9__. Its site of action is the __10__. ADH makes the distal tubule and collecting duct more permeable to __11__, thereby allowing the body to reabsorb more water from the filtrate when it is present.

1. _____

2. _____

3. _____

4. _____

5. _____

6. _____

7. _____

8. _____

9. _____

10. _____

11. _____

12. Match the term listed in column B with the descriptive phrase in column A. (There may be more than one correct answer.)

Column A

_____1. causes production of dilute urine

_____2. results in increased sodium loss

_____3. causes the body to retain more potassium

_____4. will cause water retention due to sodium movement

_____5. causes water reabsorption due to increased membrane permeability

_____6. increases sodium reabsorption

Column B

a. increased ADH

b. increased alsostrene

c. decreased ADH

d. decreased aldostrene